Governmental Interventions, Social Needs, and the Management of U.S. Forests

CONTRIBUTORS

DAVID A. ANDERSON, Forest Service, U.S. Department of Agriculture

CLARK S. BINKLEY, School of Forestry and Environmental Studies, Yale University

FRED P. BOSSELMAN, Ross, Hardies, O'Keefe, Babcock and Parsons, Chicago

MICHAEL D. BOWES, Resources for the Future, Inc.

STERLING BRUBAKER, Resources for the Future, Inc.

MARION CLAWSON, Resources for the Future, Inc.

HANS GREGERSEN, College of Forestry, University of Minnesota

PERRY R. HAGENSTEIN, Consultant, Wayland, Massachusetts

M. BRUCE JOHNSON, Pacific Institute for Public Policy Research

JOHN V. KRUTILLA, Resources for the Future, Inc.

DOUGLAS R. LEISZ, Forest Service, U.S. Department of Agriculture

BRUCE R. LIPPKE, Weyerhaeuser Company, Tacoma, Washington

WILLIAM J. MOSHOFSKY, Georgia-Pacific Corporation, Portland, Oregon

ROGER A. SEDJO, Resources for the Future, Inc.

RICHARD STROUP, Department of Economics, University of Montana

HENRY J. VAUX, Professor Emeritus of Forestry, University of California, Berkeley; Chairman, California State Board of Forestry

JOHN L. WALKER, Simpson Timber Company, Seattle, Washington

ELIZABETH A. WILMAN, Resources for the Future, Inc.

Governmental Interventions, Social Needs, and the Management of U.S. Forests

ROGER A. SEDJO, Editor

RESOURCES FOR THE FUTURE / WASHINGTON, D.C.

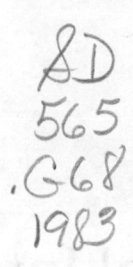

Library of Congress Cataloging in Publication Data
Main entry under title:

Government interventions, social needs, and the
management of U.S. forests.

Based on a conference held in Washington, D.C.,
March 30–31, 1981 which was sponsored by Resources for
the Future.
  1. Forest policy—United States—Congresses.
2. Forest management—United States—Congresses.
3. Forests and forestry—Social aspects—United States—
Congresses.   I. Sedjo, Roger A.   II. Resources for the Future.
SD565.G68   1983   333.75′0973   83-2879
ISBN 0-8018-3034-6

Copyright © 1983 by Resources for the Future, Inc.

Distributed by The Johns Hopkins University Press, Baltimore, Maryland 21218
Manufactured in the United States of America

Published March 1983

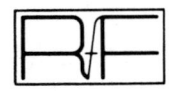

RESOURCES FOR THE FUTURE, INC.
1755 Massachusetts Avenue, N.W., Washington, D.C. 20036

DIRECTORS

M. Gordon Wolman, *Chairman*

Charles E. Bishop
Roberto de O. Campos
Anne P. Carter
Emery N. Castle
William T. Creson
Jerry D. Geist
David S. R. Leighton
Franklin A. Lindsay
Vincent E. McKelvey

Richard W. Manderbach
Laurence I. Moss
Mrs. Oscar M. Ruebhausen
Leopoldo Solís
Janez Stanovnik
Carl H. Stoltenberg
Russell E. Train
Robert M. White
Franklin H. Williams

HONORARY DIRECTORS

Horace M. Albright
Edward J. Cleary
Hugh L. Keenleyside

Edward S. Mason
William S. Paley
John W Vanderwilt

OFFICERS

Emery N. Castle, *President*
Edward F. Hand, *Secretary-Treasurer*

Resources for the Future is a nonprofit organization for research and education in the development, conservation, and use of natural resources, including the quality of the environment. It was established in 1952 with the cooperation of the Ford Foundation. Grants for research are accepted from government and private sources only on the condition that RFF shall be solely responsible for the conduct of the research and free to make its results available to the public. Most of the work of Resources for the Future is carried out by its resident staff; part is supported by grants to universities and other nonprofit organizations. Unless otherwise stated, interpretations and conclusions in RFF publications are those of the authors; the organization takes responsibility for the selection of significant subjects for study, the competence of the researchers, and their freedom of inquiry.

This Research Paper is a product of RFF's Renewable Resources Division, Kenneth D. Frederick, director. Research Papers are intended to provide prompt distribution of research having a narrower focus or a greater technical nature than RFF books.

CONTENTS

Preface	xi
Acknowledgments	xii
Introduction	xv

PART I

U.S. Forestry and Forest Land Use Issues Within National and Global Contexts

An Overview of Part I	1
1. Forest Land Use in the United States: A Review of the Past and a Look at the Future Marion Clawson	5
Comments Bruce Lippke	50
2. The Potential of U.S. Forest Lands in the World Context Roger Sedjo	53
Comments Hans Gregersen	78

PART II

Public Intervention on Private Forest Lands

An Overview of Part II	87
3. Land Use Concepts Sterling Brubaker	95
Comments Richard Stroup	115
4. State Interventions on Private Forests in California Henry J. Vaux	124
Comments William J. Moshofsky	169
5. Regulation of Private Forest Lands: The Taking Issue M. Bruce Johnson	176
Comments Fred P. Bosselman	195

PART III

Management of Public Forest Lands

An Overview of Part III   201

6. National Forest System Planning and Management:
    An Analytical Review and Suggested Approach
    John V. Krutilla, Michael D. Bowes, and Eliza-
    beth A. Wilman   207

    Comments   Clark S. Binkley   237

7. Impacts of the RPA/NFMA Planning Process on
    Management and Planning in the Forest Service
    Douglas R. Leisz   245

    Comments   Perry R. Hagenstein   258

8. National Forest Planning: An Economic
    Critique   John L. Walker   263

    Comments   David A. Anderson   297

Tables

1-1 Area of Commercial Timberland, by Ownership Class,
     in the United States, the South and the West, 1952,
     1962, 1970, and 1977   16

1-2 Volume of Growing Stock per Acre, All Species, Com-
     merical Forests, by Major Ownership Category, 1952,
     1962, 1970, and 1977   20

1-3 Percentage of Commercial Forest Area, by Stand-
     size Class and by Major Ownership Group, United
     States as a Whole, 1977, 1970, 1962, and 1952   24

1-4 Net Annual Growth per Acre of All Species, Growing
     Stock and Sawtimber, by Major Ownership Group,
     1952, 1962, 1970, and 1977   26

1-5 Estimated Productive (Wood-growing) Capacity, For-
     ests of Different Ownership Categories, for Site
     Classes I to V and I to IV, 1970   28

1-6 Annual Present and Alternative Future Growth of
     Potentially Usable Industrial Wood and of Total
     Tree Biomass in the United States   40

1-7 Average Annual Harvest of National Forest Re-
     sources, 1925-29 and 1968-72   45

2-1	Land Area of World Forest Resources, by Region and Type	54
2-2	World Growing Stock: Volume in Closed Forest	55
2-3	World Industrial Roundwood Production, 1959 and 1976	56
2-4	World Production of Industrial Conifer Roundwood, 1959 to 1956	57
2-5	World Roundwood Production, 1969 and 1976	59
2-6	U.S. Forest Products Trade for Selected Years, 1950 to 1976	61
2-7	Area of Man-made Forests, Mid-1970s, by Economic Class and Region	66
2-8	Industrial Plantations--Tropical and Subtropical America, Africa, and Asia	67
2-9	U. S. Outlook to Year 2000 for Wood Products: Softwood	72
2-10	U.S. Outlook to Year 2000 for Wood Products: Hardwood	72
4-1	Commercial and Noncommercial Forest Land in California, by Class of Ownership, 1978	127
4-2	State of California Budget Provisions for Intervention in Private Commercial Forest Resource Management: FY 1979-81	129
4-3	Timber Removal in California by Ownership, 1947-78	131
4-4	Area Reported As Planted or Seeded, by Class of Private Ownership, 1974-79	148
4-5	Estimate of Effects of State Intervention on Costs of Timber Management for Three Types of Private Owners in California	157
5-1	Index for 28 Firms and 13 Dummy Years Using Douglas Fir Prices	190
C-6-1	Present Net Worth and Losses of Forest Output Under Three Alternative Plans for Two Possible Futures	241

Figures

1-1	Total forest land, commercial forests, and cropland in farms in the United States, 1800 to 1975	6
1-2	All land in farms, improved land in farms, and commercial forest land in the South, 1850 to 1975	8
1-3	Cropland in farms and two measures of farm output in the United States, 1800 to 1980	9
1-4	Improved land in farms in the South and farm output index in the Southeast, 1900 to 1980	11
1-5	Commerical forest area and annual net growth of growing stock in all forests of the United States, 1930 to 1980	12
1-6	Commercial forest area and annual net growth of growing stock, all forest in the South, 1930 to 1980	13
1-7	Growing stock (all species) per acre, nonindustrial private forests and forest industry forests, by states, 1977	22
1-8	Growing stock (all species) per acre, nonindustrial private forests and forest industry forests, 1952, 1962, 1970, and 1977	23
1-9	Price-growth relations for Southern pine, 1952 to 1977, nonindustrial private forests	36
1-10	Price-growth relations for Pacific Coast, 1952 to 1977, nonindustrial private forests	38
1-11	Actual housing starts, softwood lumber production, softwood lumber stocks, and price of Douglas fir lumber, Housing Cycle # 5, November 1960 to November 1966	43
5-1	Index (forest prices/S&P index) of real value for hypothetical portfolio of twenty-eight forest and paper company stocks	156
5-2	Time series for Douglas fir and Southern pine prices (deflated by Wholesale Price Index), 1966-79	186

PREFACE

This book is an outgrowth of the March 1981 conference[1] sponsored by Resources for the Future at which about 100 participants from the forestry and land use communities were invited to consider critical policy issues in U.S. forestry. Attending the conference were representatives from federal and state government agencies, universities, industry, foundations, and conservation and environmental organizations.

In staging the conference, RFF sought to provide timely leadership by having some of the nation's most knowledgeable and directly concerned individuals discuss national and worldwide issues affecting U.S. forestry and forest land use.

Throughout the papers and discussants' comments presented at the conference, and included in this volume, the overriding concern involves the management of forest lands, both private and public, in a manner that would maximize the net contribution of forests to societal welfare. The issues are both practical and philosophical. The former include the performance of Forest Service planning, effects of state and federal regulation on private forest investments and management, questions of the long-run adequacy of timber resources, and outright challenges to public ownership and management as reflected in the "sagebrush rebellion." The latter are particularly timely in light of the philosophical underpinnings of the Reagan

---

[1] "Coping with Pressures on U.S. Forest Lands," March 30-31, 1981, Washington, D.C.

administration, the recent resurgence of questions concerning the appropriateness of government regulation in general and of private lands in particular, the apparent increased disillusionment of growing segments of the populace and intellectual community with regulation as currently practiced, and the continuing controversy over the legal implications of regulation as it affects asset value.

The insights provided by the conference participants are expected to be of value in helping both government and nongovernment managers and administrators increase their understanding of the issues that need to be addressed if the nation is to derive the greatest benefit from its forest lands.

The topics covered here should also be of interest to both the forestry and land use communities, as well as to economists, planners, political scientists, environmentalists, and many other groups and disciplines concerned with land use and forest management issues. The volume should be valuable as a text or as readings for university classes in forest policy.

Roger A. Sedjo

ACKNOWLEDGMENTS

Numerous individuals and organizations cooperated in the RFF-sponsored conference,"Coping with Pressures on U.S. Forest Lands," that provided many of the papers for this volume. Both the U.S. Forest Service and the Weyerhaeuser Company Foundation provided major support for the conference through their support of the Forest Economics and Policy Program of Resources for the Future. In addition, the General Services Foundation provided a major grant for the conference. Within RFF Emery Castle and Kenneth Frederick provided critical support to the undertaking of the conference.

Of course this volume would not have been possible without the contributions of the various speakers and discussants. In addition, the efforts of the session moderators, Bruce Lippke, John Barber, Carl Reidel, and Joseph Fisher, deserve a special acknowledgement.

In addition, Sally Skillings and Ruth Haas provided important suggestions as to the format of this volume, and a very special thanks to Dorothy Sawicki for her substantial editing efforts associated with the completion of this manuscript.

Finally, Lorraine Van Dine deserves a special acknowledgement for undertaking much of the preparatory work associated with the conference as well as for her efforts in typing parts of this manuscript. Also, the efforts of Cynthia Stokes, Maybelle Frashure, and John Mankin in typing the manuscript were greatly appreciated.

Roger A. Sedjo

INTRODUCTION

Throughout most of the nation's history, the demand for wood and pulp products was met largely by cutting the vast inventories of old-growth timber stands, while demand for the noncommodity outputs of the forest was light relative to its supply. Over the years demand for wood grew and recourse to more remote or less accessible timber has for many decades forced up real timber prices and permitted the use of second- or third-growth timber in the East and South. Imports from Canada and from tropical areas have relieved some of the pressure on domestic output and price, but the shift from timber mining to timber farming, with its attendant implications for supply and price, is well under way.

As indicated in the text of this volume, forest land issues are uncommonly complex. One complicating factor is the diversity of ownership: the public owns more than 25 percent of U.S. commercial forest land and the forest industry, 13 to 14 percent, while the rest—almost 60 percent—is in the hands of farmers and other nonindustrial owners. Moreover, the various ownership categories commonly have quite different objectives in managing their land. The multiproduct character of forest output is a further complication. Both timber and noncommodity forest outputs, for example, recreation, wildlife, wilderness, and watershed, all clearly have social values. Public forest managers are striving to establish management criteria that properly weight all of the social objectives, and their attempts are subject to intense scrutiny and pressure from those affected. Private owners find difficulty in realizing any private return from some forest land outputs such as improved water quality and watershed protection--these are social returns that should be considered if best use is to be made of the land. Whether on public or private land, the objective should be to produce the optimum social mix and quantity of wood and noncommodity outputs. Where markets fail to do so, other instruments can be developed to compensate for market failure. However, these other instruments may also be imperfect.

The major question addressed in this volume is the degree to which the recent mix of public and private forest land ownerships, together with the types of forest management practiced on each ownership, contributes to providing the optimal social mix of wood and nonwood outputs from the forest. Subsidiary questions involve the extent to which markets and market signals provide incentives for socially optimal forest management decisions, and the extent to which private forest land owners who have incurred financial losses due to regulation ought to receive compensation.

The 1970s saw heightened public interest in the forest land issues discussed in the following chapters. Major legislation was passed: the Forest and Rangeland Renewable Resources Planning Act in 1974, as amended by the National Forest Management Act of 1976, and the Federal Lands Policy Act in 1976. This body of legislation directs public land management agencies to plan and manage public forest lands for the continuing yield of a mix of forest services.

State and local governments have assumed a more important role in land use decisions, and these decisions affect forest lands as well. The extent to which state or local government can regulate private land use is at issue both in the political arena and in the courts. The trend has been toward more government intervention that attempts to achieve some uses and some mix of outputs different from those that would flow directly from an exclusively market-driven economy.

In addition to issues raised as a result of significant new legislation, another question that has come up with increasing frequency relates to the ability of the forest resource to service adequately the growing demand for a variety of forest outputs, many of them competitive. The picture is clouded, since the various users and interest groups all appear to believe that insufficient portions of the forest resource are being utilized to meet growing demands for the outputs with which they are primarily concerned. These differences are reflected in the continuing struggles to command control over the forest resource.

In the face of these pressures and constraints, a related question arises regarding the long-term adequacy of U.S. forests to provide timber supplies. Although periodic concern has been expressed about this issue over the last century, it continues to be important. A 1980 Forest Service study of this question--<u>An Analysis of the Timber Situation in the United States, 1952-2030</u>--gives a mixed appraisal. While finding that projected timber demands on domestic forests are increasing faster than supplies are, the study also indicates that there are large opportunities for increasing and extending timber supplies.

The convergence of increasing demand for a more diverse array of forest outputs and the growing integration of the United States into world markets for forest products, together with the diversity in ownership of forest lands and new social controls over land use, all raise complex questions about forest land management.

Public regulation of private forest lands can be used as a means to influence the mix of outputs and their levels. When is regulation desirable and appropriate? If regulation involves financial loss to private forest land owners, who should bear the costs? The latter question, sometimes called the "taking issue," is examined at length in this volume.

In the public arena concern is seen in the increased activism of interest groups attempting to influence the way in which public lands are managed. Since these lands are "public," advocacy groups believe that they have a right, or perhaps better, a responsibility, to try to influence management and policy. In addition, controversies exist as to the nature of the appropriate mix of timber and nontimber outputs to be produced on public lands. Differences in perceptions about the optimal mix of outputs generate differences in views regarding the appropriateness of various forest management regimes. The issues here relate to public policy with respect to public forest lands, the articulation of that policy in specific regulations, and the implementation of those regulations as reflected in actual management practices and procedures.

PLAN OF THE BOOK

The text is divided into three parts. Part I provides background in the form of a survey of U.S. forest resources, their potentials, some U.S. institutions, the role of this nation's forestry within the context of the world forestry community, and the dynamics of the changing nature of the forest resource. Part II addresses the issue of public intervention in the management of private lands. A rationale for government intervention is presented, and government's performance as a regulator in the State of California is examined, with the social desirability of these results being severely challenged by some of the authors. Part III introduces the objectives, difficulties, and techniques of the management of public forest lands, and discusses planning issues involving national forests.

PART I

U.S. FORESTRY AND FOREST LAND USE ISSUES
WITHIN NATIONAL AND GLOBAL CONTEXTS

OVERVIEW OF PART I

What is the state of U.S. forestry? What do we know about timber volumes, growth, ownership patterns, nontimber values, and how these affect long-run potential? What is the U.S. position as a timber producer in a global context: How is the nature of the forest resource changing, and what might one say about the U.S. ability to compete as a timber producer in the long run? These questions are addressed as follows.

The next two chapters, Marion Clawson and Roger Sedjo take a broad look at forestry and forest land use issues. Clawson's focus is on experience within the United States, while Sedjo examines the United States in a global context, looking at timber-growing potential both within this country and worldwide. Both chapters present relatively optimistic views of the possibilities and potential of the United States and the world to expand dramatically their timber-producing ability in a manner analogous to the demonstrated productivity increases experienced in agriculture over the last fifty years. This optimism is directly related to the possibilities that both authors see for improved economic forest management on both public and private U.S. forest lands.

In chapter 1, Clawson presents a historical view of forest land use in the United States, focusing on interactions between forest and agricultural land use in the South, and discusses the impact of increasing agricultural and timber-growing productivity. Both agriculture and forestry are shown to have experienced substantial gains in output and annual growth without major additions to the land areas devoted to these activities. The shifts of land between agricultural and forestry uses that did occur are attributed primarily to economic forces.

Clawson also examines ownership patterns, timber stocking, timber growth, wood growth potential, and nonwood forest outputs in the United States. He shows that, while diverse, the forest land ownership pattern in the United States has remained quited stable since 1952, with the major changes being the reduction in farmer-owned forests and an increase in a catch-all category called "other private." Furthermore, the data show that between 1952 and 1977, growing stock per acre (timber inventory) for all commercial forests in the United States increased about 25 percent. Associated with this increased timber inventory was an expansion in timber growth. The annual net growth per acre increased 64 percent for all comercial forests in the United States between 1952 and 1977. Clawson then reviews several studies that estimate the possible and probable annual growth of wood fiber and the relation of output to prices, concluding that the "ultimate potential annual growth" of potentially usable industrial wood is 60 billion to 100 billion cubic feet. This compares with the 1977 harvest of 21.7 billion cubic feet. The biomass potential he estimates at about twice that of the industrial wood.

Clawson concludes that economic forces will result in increased incentives for efficient management and in the realization of some of the currently unused potential of the forests to produce both wood and most nonwood outputs. Thus, he believes that U.S. output will expand to meet the growing demands.

In his comments on chapter 1, Bruce Lippke emphasizes the importance of technology in the future of the wood-producing industry and relates this to the adequacy of markets. He challenges some of the Forest Service projections referred to by both Clawson and Sedjo.

In chapter 2, Sedjo expands the analysis to include a worldwide perspective. After briefly discussing the existing world forest resource base and the current position of the United States within that global context, the author examines the transition in the nature of the forest resource from old-growth to intensively managed forest plantations and compares it with the earlier transition experienced in meeting food needs as mankind moved from reliance on hunting and foraging to livestock raising and cropping. Initially, forest resources were very low. However, as the old growth diminished and demand increased, real stumpage prices rose, reflecting the growing economic scarcity of timber. The higher stumpage prices provided incentives for investments in intensive forest management, man-made forests, and tree-growing technology. Thus, Sedjo believes that economic forces will generate incentives for increased forest management.

Furthermore, as forest location becomes a decision variable, the location of plantation forests can be expected to gravitate to sites with high biological and economic potential. Future industrial wood requirements, according to Sedjo, are likely to continue to be met from a variety of sources, but the supply of industrial wood will increasingly be provided by high-yielding forest plantations rather than by old-growth stands.

Sedjo estimates that high-yielding, man-made forests on only a small portion of the world's current forest lands could, potentially, easily meet the world's projected industrial wood requirements into the twenty-first century. He also estimates that the high-yielding sites in the United States alone would be capable of producing a substantial portion of projected world demand into the twenty-first century.

In his comments on chapter 2, Hans Gregersen points out a number of real-world problems related to the potential role of tropical and Third World countries becoming major sources of forest products in the next twenty to fifty years.

First, Gregersen cautions that demand for wood may grow more rapidly than currently envisioned, particularly if fuelwood becomes an increasingly important source of energy. Demands on forests and forest lands for fuelwood production could divert substantial resources away from the production of industrial wood both in the United States and elsewhere. Second, large amounts of capital and technical/managerial expertise will be required before plantations can become an important factor in the totality of worldwide production and trade. While not suggesting that these forces will preclude the type of scenario envisioned by Sedjo, Gregersen cites these as "dampening" factors.

Gregersen goes on to note the important role being played by technology and international technology transfers. Examples cited indicate the difficulty in projecting future demand, especially for a particular species or type of fiber. Technology is bound to result over time in substantial shifts in demand for various wood types and hence to affect the pressures being brought to bear on particular forest resources.

In summary, the introductory chapters provide a broad description of some salient features of the forests of the United States and speculate as to their long-term potential for timber production. The conclusions tend to be optimistic with respect to the potential of both U.S. and world forests to expand timber production dramatically over the longer term if the appropriate management decisions are made for both private and public lands. While both chapters exhibit faith in the ability of management to respond to economic incentives, a Pollyannaish view is deemed inappropriate, given the numerous real-world obstacles that must be successfully negotiated before the potential can be realized.

Chapter 1

FOREST LAND USE IN THE UNITED STATES:
A REVIEW OF THE PAST AND A LOOK AT THE FUTURE

Marion Clawson

A conscious and wise strategy for the future use of forest lands, as for any other natural resource program, requires the best possible understanding of the past and the present. The future will certainly be different from the past, quite probably in presently unimaginable ways, yet the past is the most reliable guide to the future if that past is properly understood and interpreted.

When the explorers and colonists from European countries first landed in eastern North America, most of the land was covered with dense forest. Those forests were enormously impressive to the first colonists, because of the variety, large size, and high quality of their trees. More than 40 percent of what is now the contiguous forty-eight states of the United States was covered with what today is officially described as "commercial" forest, and additional areas had considerable tree cover, but at lower densities or with lower potential growth rates or both (figure 1-1). Land clearing began early, and through the nineteenth century and the first quarter of the twentieth century, a large part of this forested area was cleared for crop agriculture. Some land was cleared, farmed, abandoned, grew up to forest, and was cleared again, in numerous and complex forest land histories. Some of the forest was cleared for towns and cities, and other forests were cleared for roadways and other purposes, but crop agriculture has been the major competitor with forestry for the use of potential forest land. Grazing on native grasses and other forage plants occupies a very large area also, but most of the grazing land never supported what we now call commercial forest.

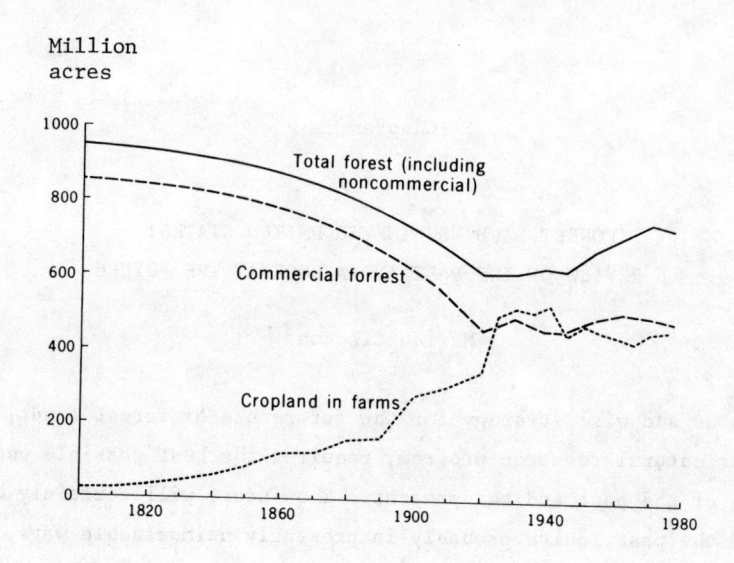

Figure 1-1. Total forest land, commercial forests, and cropland in farms in the United States, 1800 to 1975

Source: Marion Clawson, "Forests in the Long Sweep of American History," Science vol. 204 (June 15, 1979) p. 1169. Copyright 1979 by the American Association for the Advancement of Science.

In a general way, this history of forest land use on a national scale is reflected in the history of forest land use in the South (figure 1-2). Although some settlement began in the South at an early date by American land settlement standards, the area of land cleared for crop agriculture remained comparatively small until the last quarter of the nineteenth century. During that period and the first quarter of the twentieth century, land clearing proceeded rapidly, and improved land in farms reached a peak in the 1930s, from which it has receded by nearly a third. As in other areas, some forest was cleared, the land was farmed, and then reverted to trees, perhaps to be cleared again and to revert again.

Both nationally and in the South, a considerable stability in the cropland/forest land balance has been reached in the past thirty to fifty years. That is, the total acreages of each land use in each region have ceased to change at the rate they changed earlier. Some land, both forested and cropland, continues to be taken for other uses, including urban expansion, highway rights-of-way, power line rights-of-way, and reservoir uses, but the massive land use changes associated with the westward tide of settlement of the United States are matters of the past. As Held, Stoddard, and I said more than twenty years ago, the net changes in land use in the future will be relatively small, but the changes that do occur will come with increasing "turmoil" or controversy (Clawson, Held, and Stoddard, 1960).

Shifts in area of land used for different purposes are important, but are only part of the story for both agriculture and forestry. On a national scale, agricultural output expanded more or less proportionately to the increase in cropland area throughout the nineteenth century and for about the first half of the twentieth century (figure 1-3). The dramatic and significant fact shown on figure 1-3 is the substantial increase in farm output from the time after the 1930s to the present, with little change in cropland area. Output has essentially doubled from the same or a slightly declining cropland area. Clearly, cropland use has become more intensive over this generation of agricultural land use, with the margin of agricultural output at the intensive, not at the extensive, margin of land use.

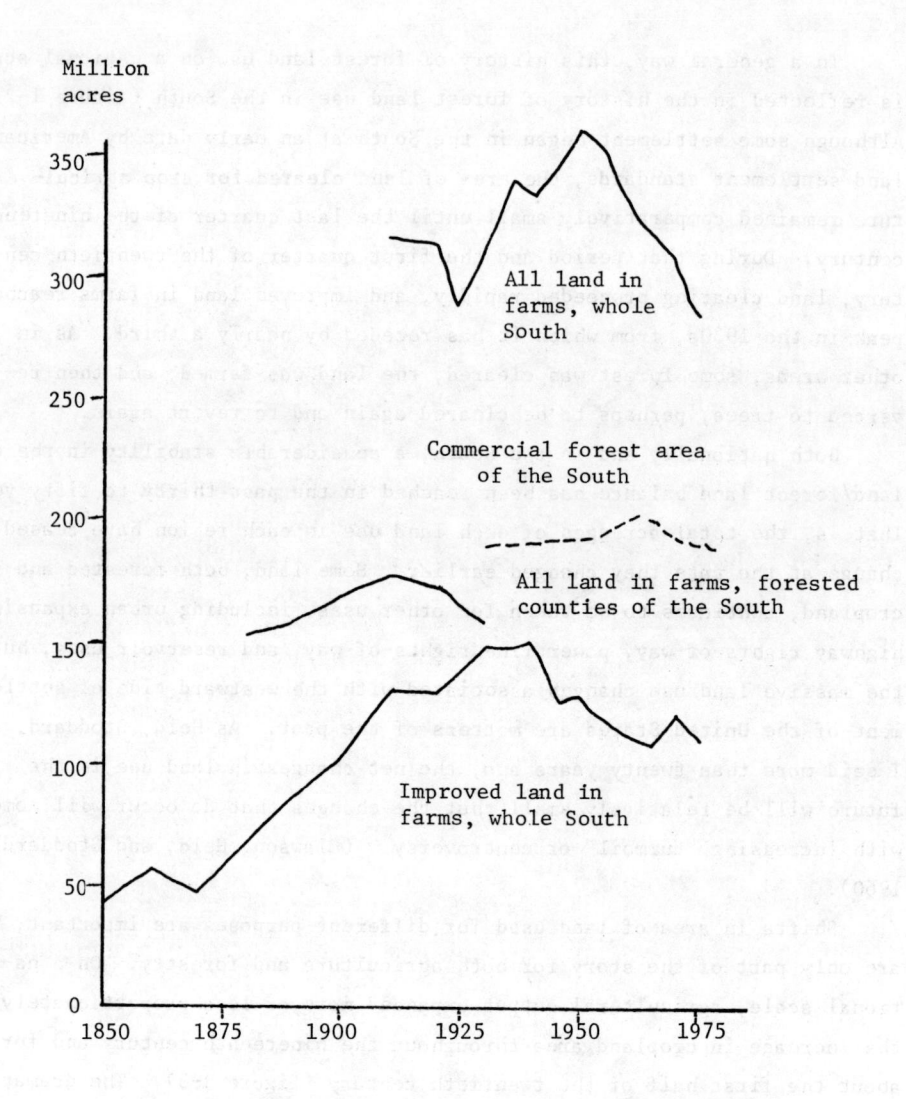

Figure 1-2. All land in farms, improved land in farms, and commercial forest land in the South, 1850 to 1975

Source: Adapted from Marion Clawson, *The Economics of U.S. Nonindustrial Private Forests* (Washington, D.C., Resources for the Future, 1979) pp. 72, 336.

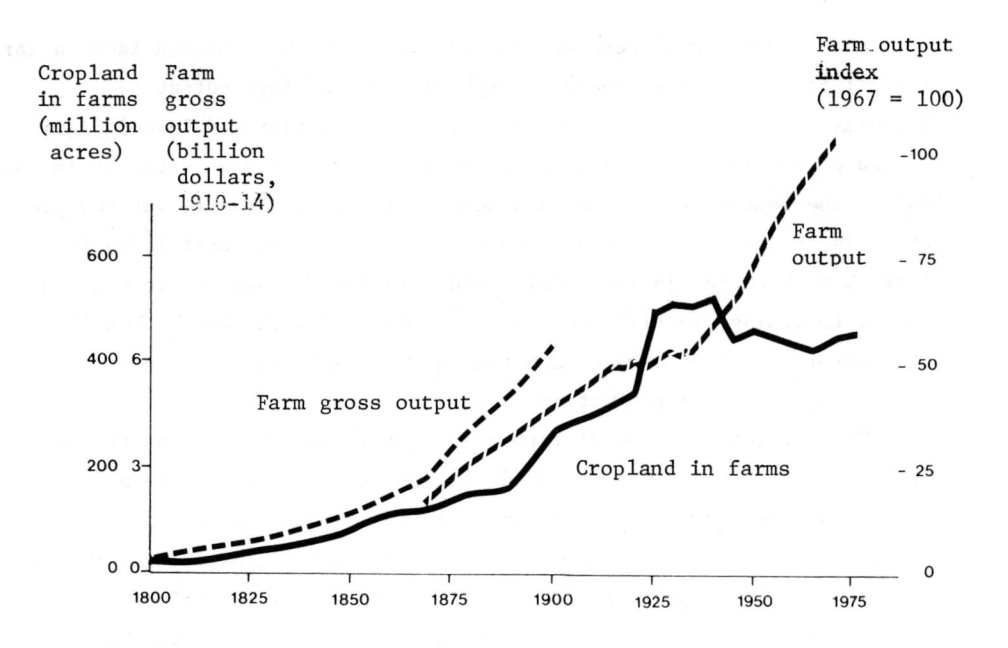

Figure 1-3. Cropland in farms and two measures of farm output in the United States, 1800 to 1980

Source: Marion Clawson, "Competitive Land Use in American Forestry and Agriculture," *Journal of Forest History* vol. 25, no. 4 (1981) p. 224.

A very similar picture emerges when the area of improved land in farms for all of the South is compared with an index of farm output for the Southeast (figure 1-4). As shown in figure 1-2, improved land in farms peaked in the 1920s and since then has declined nearly a third. Farm output in the Southeast has varied somewhat from year to year, but the overall trend is steeply upward, approximately doubling in the past fifty years. There have been considerable differences in trends, both as to magnitude and as to timing, in different parts or regions of the South, but the overall picture is clear enough from figure 2-4--there has been an intensification of agricultural production.

The national picture of farm output trend in relation to cropland area is closely paralleled in the trends in forest production nationally (figure 1-5). As noted earlier, forest area decline throughout the nineteenth and early twentieth centuries, and has remained roughly constant for the past half century or slightly longer. In the mature forests that the colonists found, net growth of wood was zero or close to it; the growth that did take place was offset by mortality due to fire, storm, disease, insects, old age, and rot. As the original forests were cut, some of that land went into agricultural crop production, as noted; the land not cropped, or cropped and later abandoned for cropping, came back to trees, but slowly, and often the species were different from those in the original forest. There was a common expectation that the land cleared of forest would go into production, and forest clearing was generally regarded as a constructive step. Fires were often set to prevent forest regrowth, and little or no effort was made to fight the forest fires that did occur. As a result, net growth of wood was low until after the end of the first quarter of the twentieth century. Since then, annual growth of wood has increased greatly, by about 3 1/2 times from 1920 to 1977. As with crop agriculture, the margin of forest output has been intensive, not extensive.

The same general forest picture is found for the South (figure 1-6). As noted earlier, commercial forest area has been very nearly constant since about 1930, with additions to forest on land abandoned for crop agriculture about balancing loss of forest land to agriculture and other land uses. The net annual growth of wood has nearly doubled in the past fifty years. There have been increases in net growth of wood for both hardwoods and softwoods.

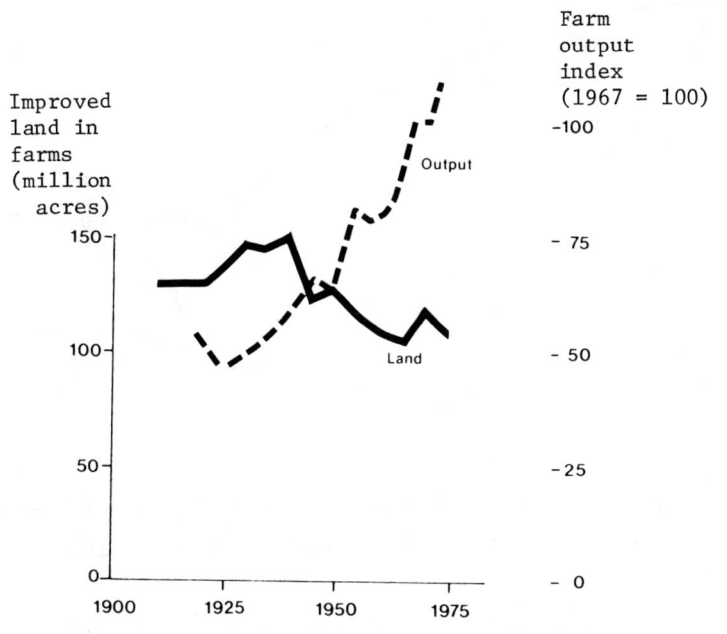

Figure 1-4. Improved land in farms in the South and farm output index in the Southeast, 1900 to 1980

Source: Marion Clawson, "Competitive Land Use in American Forestry and Agriculture," Journal of Forest History vol. 25, no. 4 (1981) p. 224.

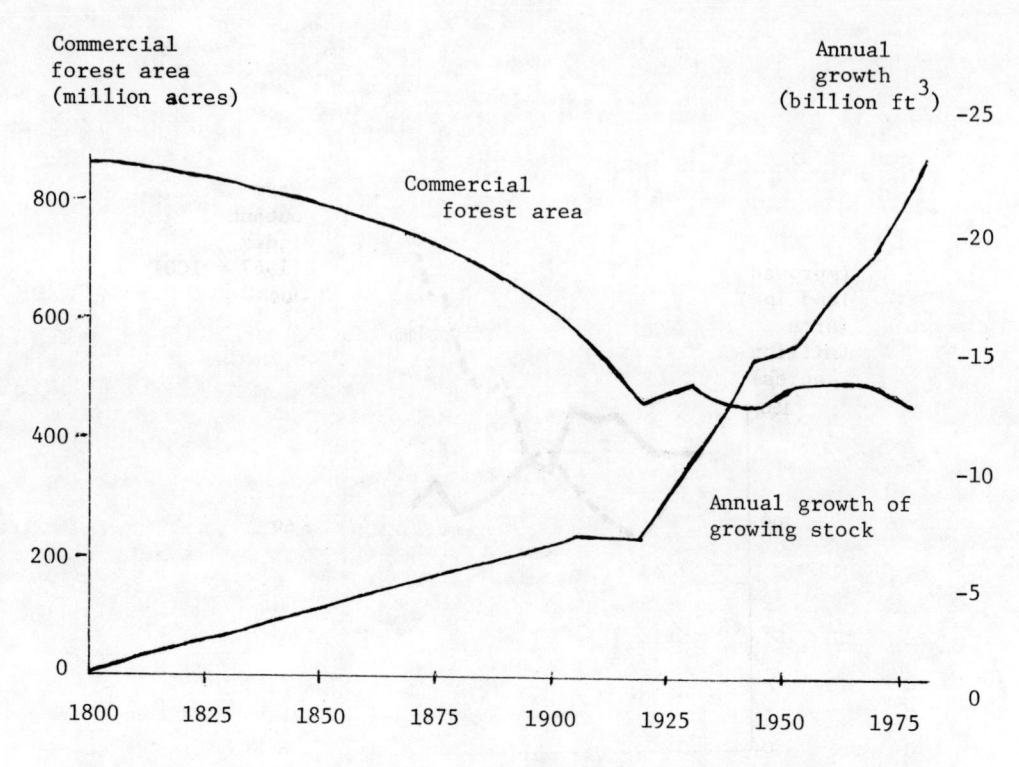

Figure 1-5. Commercial forest area and annual net growth of growing stock in all forests of the United States, 1800 to 1980

Source: Adapted from Marion Clawson, "Forests in the Long Sweep of American History," Science vol. 204 (June 15, 1979).

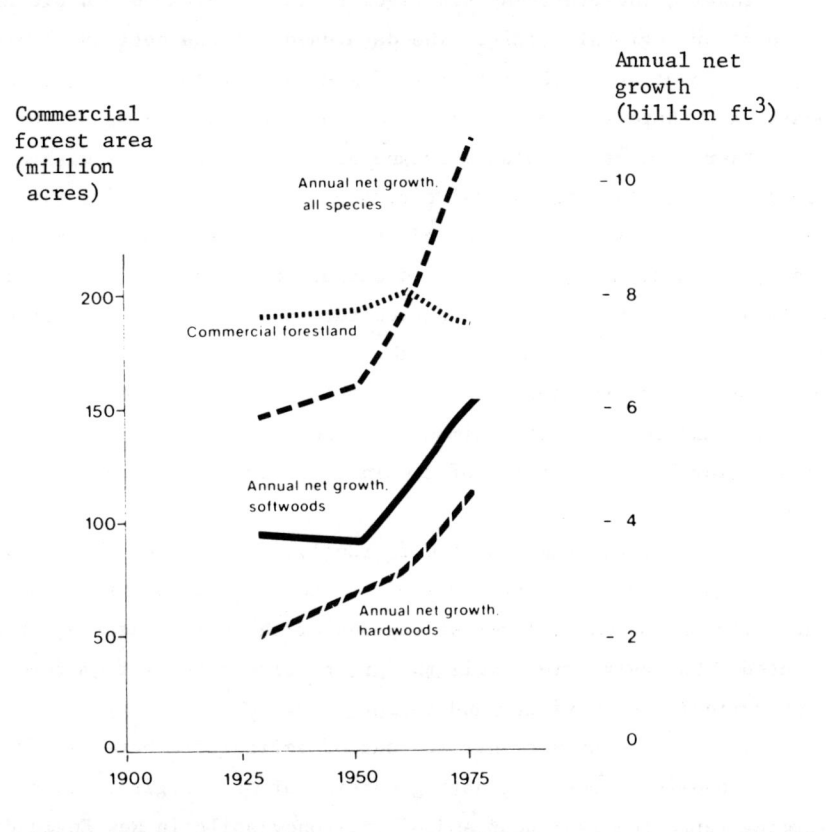

Figure 1-6. Commercial forest area and annual net growth of growing stock, all forests in the South, 1930 to 1980

Source: Marion Clawson, "Competitive Land Use in American Forestry and Agriculture," *Journal of Forest History* vol. 25, no. 4 (1981) p. 225.

Clearly, the preceding six figures present broad-brush pictures of national or regional totals. The data used are the best available, but there is doubt as to the accuracy of some data, especially for the early years, and there are doubts as to the comparability of data from one period to another. There are also substantial local differences within these large totals. But the general picture--of an earlier major shift in land use followed by a large degree of stability in land use for the past half century, and of major increases in output from more or less constant areas of land--is accurate for both agriculture and forestry, and for both the nation as a whole and for the South.

Some generalizations can be drawn from these figures:

a. Agriculture has claimed land and has relinquished it for other uses primarily on the basis of the income returns from agricultural production.

b. To put the same point differently, forestry has been a weak competitor for land. When agriculture relinquishes land, trees come back naturally in the natural forest regions of the United States, or they are planted. But only rarely will the prospective returns from forestry force crop agriculture to yield land to forestry.

c. The land shifts into and out of agriculture have been due primarily to economic forces originating outside of each region. Thus, crop farming retreated from poor and often stoney soils in New England as agricultural expansion into the Midwest brought agricultural prices down to levels where the returns from New England farming on some lands were too low for farmers to continue. Cotton left the hilly lands of the Piedmont of the Southeast as cotton acreage expanded on the High Plains and in the western irrigated areas. And so on.

d. As agricultural area shrank in any region or state, ordinarily the lands least productive for crop agriculture were given up (although least productive within their region or state, these might not have been the least productive lands in cropping when viewed nationally). Some of the lands given up by crop agriculture have formed relatively productive bases for forestry, which is particularly true in much of the South.

e. When crop agriculture has expanded in any region or state, by and large it has claimed the most productive sites. However, some of the shifts to cropping have been tempered by the costs of clearing the land from forests, and sometimes these costs have been higher on the best land than on only moderately productive land.

OWNERSHIP OF FOREST LAND

Not only has the total area of commercial forest land been relatively stable since about 1920, but the ownership situation has also been relatively stable since 1952 when data on forest land ownership began to be available and reasonably comparable from area to area and from one date to another (table 1-1). There was a modest increase in commercial forest area from 1952 to 1962 for the United States as a whole, in the South, and in the West; since then, there have been modest declines in each, somewhat more marked since 1970 than previously. In this relatively stable land use situation, the forest industry firms have expanded their holdings by about 12 percent nationally and by almost the same proportion in the South.

The most marked trend shown in table 1-1 is the sharp decline in farmer-owned forest land, especially nationally and in the South, and to a lesser degree in the West, and the largely offsetting rise of the "other private" class.[1] (The "farmer" ownership class represents forest land owned by farmers, regardless of whether or not the land is in an operating farm; it does not include forest land within operating farms owned by non-farmers and rented by farmers.) In the years after 1952, the number of operating farms in the United States declined by about half; while most of the land in the farms no longer operated independently has been absorbed into the remaining farms, some forested land evidently has not been taken into the remaining farms.

The "other private" in table 1-1 are highly variable, having in common only the facts that the land is privately not publicly owned; that the

---

[1] In _The Economics of U.S. Nonindustrial Private Forests_ (Clawson, 1979a), I question the accuracy of these data; see especially chapter 5. I conclude that "it is altogether probable that the acreage of farm forests as reported by the Forest Service in 1977 is too high, perhaps by a third or more" (p. 62). I also seriously question the accuracy of earlier data on farm and "other private" ownership, especially the data for 1952. But there is no doubt in my mind that the trends in forest ownership as reported in table 1-1 are generally in the right directions, even if the actual numbers are in error.

Table 1-1. Area of Commercial Timberland, by Ownership Class, in the United States, the South, and the West, 1952, 1962, 1970, and 1977

(million acres)

Ownership class	1952	1962	1970	1977
United States				
National forest	94.7	96.9	94.6	88.7
Other public	49.0	44.8	47.0	45.0
Forest industry	59.5	61.6	67.0	68.8
Farmer	172.3	144.8	125.3	115.8
Other private	123.8	159.3	162.6	162.2
Total	499.3	509.4	496.4	482.5
South				
National forest	10.4	10.7	10.8	11.0
Other public	6.4	6.5	6.7	6.7
Forest industry	32.1	33.4	35.1	36.2
Farmer	91.3	73.0	62.8	55.9
Other private	51.9	76.3	77.1	78.2
Total	192.1	199.9	192.5	188.0
West				
National forest	74.0	75.9	73.4	67.9
Other public	21.5	20.2	20.1	19.6
Forest industry	13.5	14.1	14.5	14.6
Farmer	15.9	15.4	14.5	13.9
Other private	13.5	13.0	12.7	12.3
Total	138.5	138.6	135.3	128.3

Source: U.S. Department of Agriculture, Forest Service, <u>An Analysis of the Timber Situation in the United States, 1952-2030</u>, Review Draft (Washington, D.C., 1980) app. 3, table 4, pp. 22-25

land is not owned by a farmer; and that the forest land owner lacks processing facilities to manufacture lumber or other forest products from the forest. "Other private" forests are often erroneously referred to as "small"; while indeed most of them were reported in 1952 as having fewer than 100 acres, and while this may well be true in 1977, in fact some have quite large holdings, up to half a million acres in at least one case.

Although the trends in forest land ownership from 1952 to 1977 are relatively similar in the South and in the West, the pattern of ownership differs sharply between these two major regions. In the West, national forests include 53 percent of the commercial forest land (according to the Forest Service definition of commercial; if the standards are stricter, the percentage included in national forests is less) and other publicly owned forests include additional areas, so that all public ownership is 68 percent of the total. In the South, the comparable figures are 5 percent and 9 percent. The forest industry owns 14 percent of the forest land of the nation, but 19 percent in the South and only 11 percent in the West. The farmer and "other private" groups own only 20 percent of the total in the West, but 71 percent in the South.

The general stability in ownership of commercial forests since 1952, as reported in table 1-1, is likely to continue. Some forested land may be purchased by some unit of government from private owners; this is likely to be more than offset by the reservation of publicly owned forest land for park, wilderness, and other uses that remove it from the commercial category. Forest industry firms may continue to increase their holdings as opportunities arise, but the typically high price of privately owned forest land, due in large part to recreational possibilities and also to the substantial inflationary increase in prices, will put a damper on such acquisitions. There may be some continuing decline in farmer-owned forests, but most of the adjustment in farm sizes and hence in farm numbers as a result of technological changes and economic pressures has apparently been made. There may well be continuing shifts in forest ownership within the "other private" category. But those projecting or forecasting forest production and forest land use for the next few decades would be well advised, I think, to accept something closely akin to present forest land ownership as a basic factor in their planning.

TIMBER STOCKING ON U.S. FORESTS

Forests typically involve heavy capital investment in standing timber. This is true almost regardless of whether the objective of the forest management is the production and harvest of wood or the preservation of a mature forest, and whether it is any intermediate management goal. Trees are capital on which additional wood is produced; when harvested, they are output. In this respect, a tree is similar to a heifer in a cattle herd. The optimum volume of timber in a forest, or the optimum investment in standing timber, at any date depends on several factors:

1. The age of the trees in the stand: very young trees have very small volumes and older trees, larger volumes, at least up to some old age of the trees;

2. The dominant tree species, since some species normally grow to larger sizes and larger volumes per acre than do others;

3. The management objectives of the forest owner/manager, since some forest management programs normally involve larger volumes of wood per acre than do other management programs, even for similar sites.

The actual volume of wood on any forest site, or for a class of forest sites, depends upon the foregoing factors and also upon the history of the forest stand, which in turn affects age of trees, degree of stocking, thrift of the stand, and other factors. Data on average volume of timber on a particular stand or on a class of stands (for example, an ownership group) are not easy to interpret without a great deal of data on the foregoing factors--and these data often are lacking. A specific number of cubic (or board) feet per acre may seem large, yet given the circumstances of the stand may be less than optimum; conversely, a specific number that seems low may actually be highly satisfactory. For instance, the volume per acre on a forty-year-old stand should be vastly larger than the volume on a ten-year-old stand. These caveats must be kept clearly in mind in interpreting the data that follow.

In the United States, available data on forest stands may be in terms of sawtimber, which measures 7, 9, or 11 inches or larger diameter (breast high), depending upon the area, the species, and customary industry standards; or the data may be in terms of growing stock, which includes all trees 5 inches or more in diameter, up to a 4-inch top. In either case,

the data include only species that are used for industrial wood purposes--
that is, they exclude fuel and other woods not used for lumber or other
product manufacture or not used as poles or piling. Also, in each case
the volume is measured only up to the point where the tree stem breaks
into limbs. The volume of sawtimber is measured in board feet, the volume
of growing stock in cubic feet. The growing stock definition is somewhat
more inclusive than the sawtimber definition, but each excludes trees less
than 5 inches in diamter, limbs and tops, and species not ordinarily used
for industrial wood purposes. Each is therefore substantially below total
biomass--how much below depending upon a particular situation; not infre-
quently the measured quantities are half or less of total biomass.

Between 1952 and 1977 the volume of growing stock per acre for all
commercial forests in the United States increased about 25 percent (table
1-2). The largest increase (in both absolute and relative terms) was on
the nonindustrial "other private" forests, where average stocking rose by
46 percent in twenty-five years. The national forests had in 1952, and
still have to a lesser degree, old-growth or mature timber, normally in
large volumes per acre (for the particular species and site classes).
Timber harvest of these old-growth stands results in new, younger, smaller-
volume stands (at least for some years), even when reproduction is prompt
and satisfactory, and more so if reproduction is delayed or less than optimum in volume. However, in spite of the harvest of old growth, which would
be expected to lead to lower volumes of standing timber per acre, in fact
the average stand per acre on national forests increased by 11 percent over
this period. The forest industry forests show the least variable stocking
rates of any ownership group, with a variation in stand per acre from one
enumeration date to another of less than 5 percent.

On the basis of the 1977 data, the nonindustrial private forests as a
whole had an average rate of stocking only 73 percent of that for all
forests; national forests, on the other hand, had an average rate of
stocking 175 percent of average for all forests. Is the former too low and
the latter too high by some standard? It is impossible to make a firm conclusion without further data on age of stands, species composition, and
other factors.

Table 1-2 does indicate clearly, however, that the United States is
building up its stock of standing timber on the average for each major

Table 1-2. Volume of Growing Stock per Acre, All Species, Commercial Forests, by Major Ownership Category, 1952, 1962, 1970, and 1977

(cubic feet)

Year	All	National forests	Other public	Forest industry	Other private
1952	1,175	2,270	1,325	1,585	735
1962	1,230	2,360	1,430	1,552	783
1970	1,298	2,370	1,540	1,485	891
1977	1,465	2,570	1,610	1,547	1,070

Source: Marion Clawson, The Economics of U.S. Nonindustrial Private Forests (Washington, D.C., Resources for the Future, 1979) p. 127. (Data in the source are from Forest Service reports on timber demand, supply, and projections.)

ownership category. This does not, of course, prove that stocking is ideal in any category, much less that it is ideal for every forest tract in each ownership category. But it does clearly indicate that any talk about forest depletion or forest exhaustion on a national scale is in error; we are building up, not reducing, our forest stands.

The experience on forest industry forests, as shown in table 1-2, may provide one guide to the matter of the optimum rate of stocking. This is so for two reasons: first, the average rate of stocking on industry forests has been relatively constant, at least since 1952, as noted earlier; and second, it may be assumed that forest industry firms, motivated by profit maximization, have attained something approximating an optimum economic rate of stocking for their forests as a whole. This does not assume that every forest industry tract is stocked at its optimum, but only that in general forest industry firms have more closely approached economic stocking than has any other major ownership group. When nonindustrial forests are compared state by state with forest industry forests, it appears that the nonindustrial private forests are almost as well stocked as are the forest industry forests (figure 1-7). It may be assumed that forest industry firms have sought, in general, to obtain productive forest lands, while at least some nonindustrial private forests are on less productive sites--which, however, may be equally useful for nonwood outputs such as outdoor recreation or wildlife. Some disparity between nonindustrial and forest industry forests would thus be expected.

As indicated in table 1-2, nonindustrial private forests had an average rate of stocking per acre only 69 percent as high as did forest industry forests; in figure 1-7, nonindustrial private forests have an average rate of stocking about 81 percent that of forest industry forests. Why the discrepancy? Because forest industry forests are more common in states (and in regions within states) where the productivity of all forests is high, and many nonindustrial private forests are located in states (and regions within states) where the productivity of all forests is relatively low. The analysis in figure 1-7 thus goes partway toward meeting the problems of noncomparability of forest situations briefly discussed at the beginning of this section; and on the basis of this analysis, one may say that nonindustrial private forests are not much, if any, less fully stocked than are forest industry forests under comparable conditions.

Over the years, nonindustrial private forests have made substantial gains on forest industry forests (figure 1-8). From having only about half as heavy stocking state by state in 1952, they advanced in 1977 to the 81 percent of forest industry stocking noted earlier. It may well be that many persons who describe the nonindustrial private forests as understocked are inadvertently harking back twenty-five years or more.

The matter of stocking rates on forests of different ownership categories can be approached by a consideration of the proportion of the total forest area in different stand-size groups (table 1-3). While there was a substantial stability among the stand-size groups for each ownership category in the 1952-77 period, there was some trend toward more sawtimber stands and more seedling-sapling stands and away from pole timber stands for all forests combined. Even the national forests, in spite of the liquidation of old-growth stands, had a relatively larger area in sawtimber stands in 1977 than in 1970. The national forests differ substantially from each of the other major ownership categories in their very high propor-

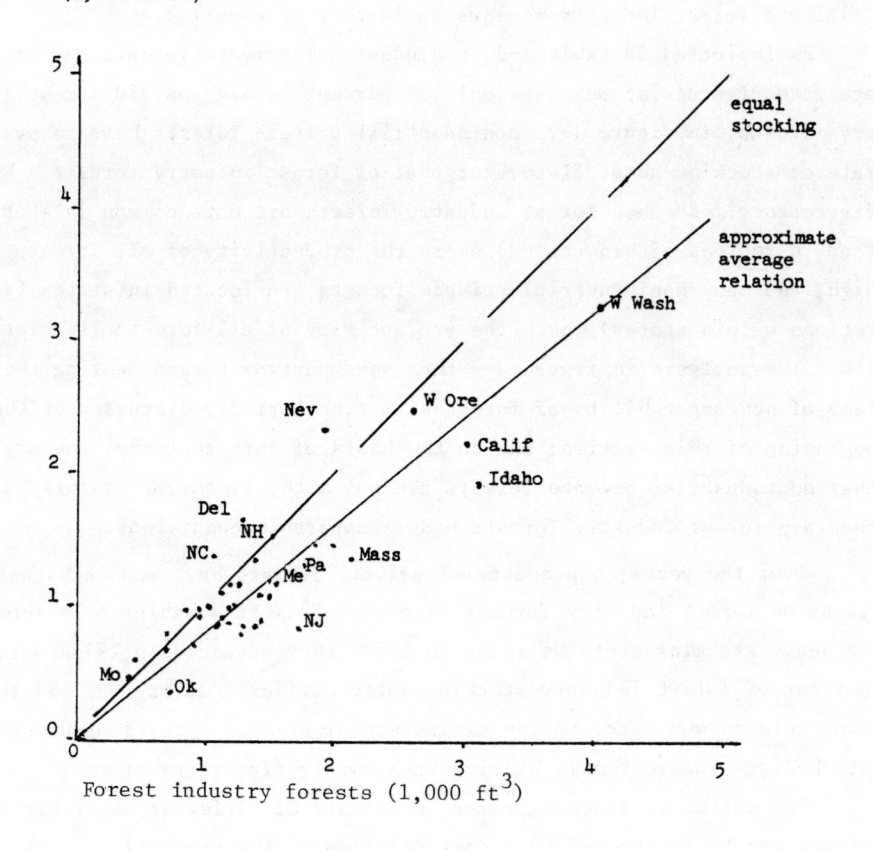

Figure 1-7. Growing stock (all species) per acre, nonindustrial private forests and forest industry forests, by states, 1977

Source: Marion Clawson, The Economics of U.S. Nonindustrial Private Forests (Washington, D.C., Resources for the Future, 1979) p. 133.

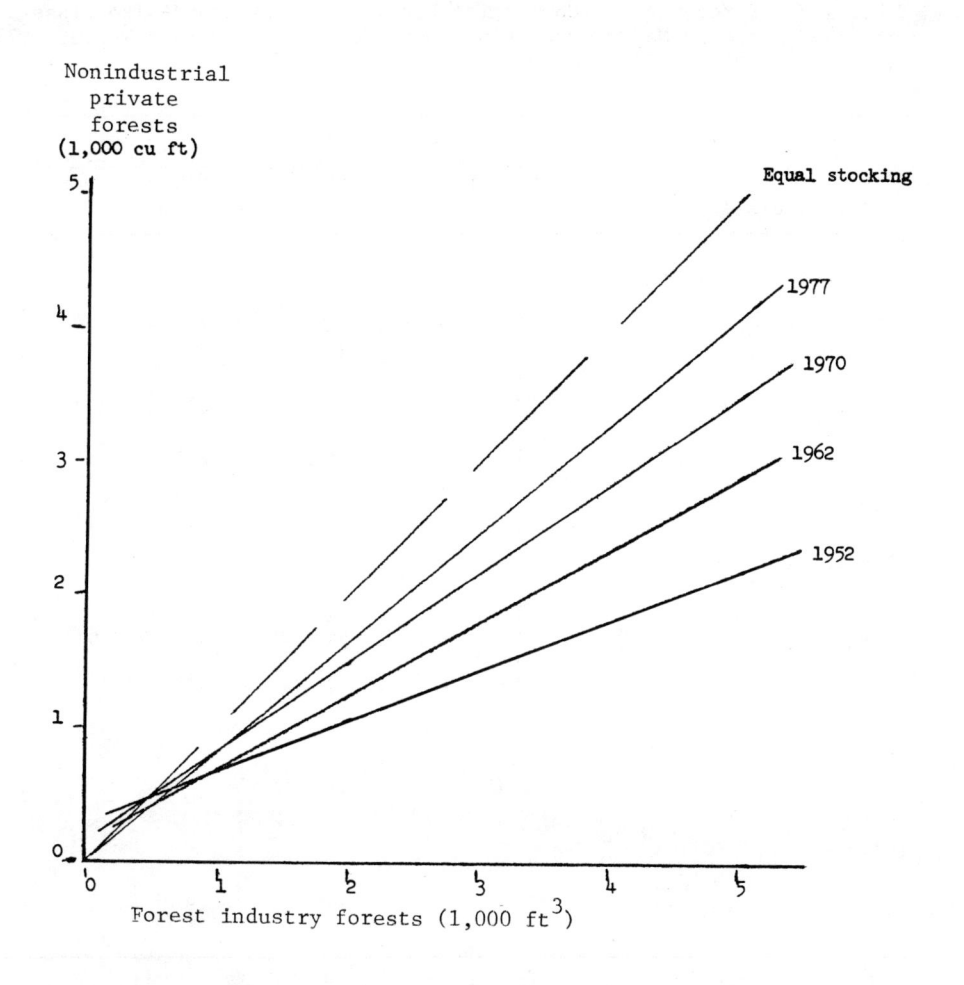

Figure 1-8. Growing stock (all species) per acre, non-industrial private forests and forest industry forests, 1952, 1962, 1970 and 1977

Source: Marion Clawson, The Economics of U.S. Nonindustrial Private Forests (Washington, D.C., Resources for the Future, 1979) p. 134.

Table 1-3. Percentage of Commercial Forest Area, by Stand-size Class and by Major Ownership Group, United States as a Whole, 1977, 1970, 1962, and 1952

Ownership class	Percent of commercial timberland area in:			
	Sawtimber stand	Pole timber stands	Seedling, sapling	Nonstocked areas
1977				
National forest	64	22	11	3
Other public	46	30	20	3
Forest industry	45	25	28	2
Other private	38	31	28	4
All	45	28	24	3
1970				
National forest[a]	59	20	12	4
Other public	42	31	21	6
Forest industry	48	22	28	2
Other private	37	27	31	4
All[a]	43	25	26	4
1962				
All	41	32	20	7
1952				
All	37	35	19	9

Sources: Various Forest Service reports on supply and demand for forest products, including statistical appendixes.

[a] Excludes 5.0 million acres in Rocky Mountain region.

tion of sawtimber stands and their consequent lower proportions in stands with smaller timber. The other ownership categories, while having some differences, are fairly similar among themselves. The situation on the national forests is partly a product of their history (the harvest of old-growth timber has been delayed in large part because of the relative remoteness of many national forest areas) and partly a product of legislation and agency timber management philosophy, which has resulted in a very slow rate of harvest of the old-growth stands.

TIMBER GROWTH, 1952 TO 1977

In a great many ways, annual net growth of timber is more important than is the volume of standing timber. Growth is output, or production, and this is the basis of income, both "real" and monetary. Stand is inventory; some inventory is indispensable, but inventory is valuable only as it is used. Within considerable limits, harvest and growth do not need to be identical, but rather can diverge. That is, timber harvest can proceed out of available inventory, but obviously only to the extent of that inventory --an attempt to cut more than grows would eventually lead to a zero inventory. Conversely, harvest may fall short of net growth for some years, with the unharvested growth simply added to inventory. But there are limits to this also, since inventory cannot exceed some maximum volume--a volume that is dependent upon the site characteristics and upon the tree species involved. Growth and inventory are necessary for timber harvest, but harvest is equally necessary for net growth beyond some level.

Between 1952 and 1977, the annual net growth (of growing stock) per acre increased 64 percent for all commercial forests in the United States (table 1-4). The rate of increase between successive periods was 11 percent, 12 percent, and 12 percent, respectively; since the length of the period between inventories decreased, the annual rate of increase rose, and at an increasing rate. In absolute terms, or cubic feet per acre, the rising trend was more marked and accelerated faster in recent years.

At each inventory date, the rate of annual net timber growth per acre was less on national forests than on forests in any other major ownership category. Over the twenty-five years from 1952 to 1977, the absolute increase in annual net growth per acre was less (12 cubic feet) for national

Table 1-4. Net Annual Growth per Acre of All Species, Growing Stock and Sawtimber, by Major Ownership Group, 1952, 1962, 1970, and 1977

	Actual growth				Potential capacity[a]
Forest ownership	1952	1962	1970	1977	
Growing stock (cubic feet)					
National forests	23	26	28	35	72
Other public	26	32	39	54	77
Forest industry	43	50	52	5/	88
Other private	27	31	34	45	74
All	28	32	37	46	75
Sawtimber (board feet)					
National forests	83	98	107	143	--
Other public	87	109	130	184	--
Forest industry	158	174	185	213	--
Other private	81	89	108	140	--
All	91	103	120	151	--

Source: Marion Clawson, The Economics of U.S. Nonindustrial Private Forests (Washington, D.C., 1979) p. 153.

Note: Dash = not applicable.

[a] In a fully stocked natural stand, at an age of approximately mean annual increment, with good but not intensive forest management; of species now on sites.

forests than for any other major ownership category. Although the forest industry forests increased their annual net growth more in absolute terms than did the national forests, this was from a much higher level of output in 1952; hence in percentage terms the industry forests did slightly poorer than the national forests. The other major ownership categories did better, both in absolute and in relative terms, than did the national forests.

The actual net wood growth per acre in 1977, while substantially higher than in earlier years, was still far below the potential from good natural forestry on these lands. Based on the figures in table 1-4, for all forests in 1977 the actual net growth per acre was 61 percent of the potential growth under good natural forestry; while this may seem low, it is clearly far above the 37 percent figure for 1952. In a comparison of actual with potential net growth, as in the earlier comparisons the national forests again come out the lowest, at 49 percent in 1977. The other public forests have the highest rate, at 70 percent, followed closely by the forest industry forests at 66 percent of potential.

The estimated potential net growth rate per acre under good natural forestry is much influenced by the Forest Service's definition of commercial forests. One critical element in that definition is the ability to grow more than 20 cubic feet of industrial wood per acre annually under specified conditions. There is good reason to believe that this threshold is too low if one is concerned with economic timber production. The Forest Service does classify commercial timberland into five productivity site classes. The least productive class has the capacity to grow from 20 to 50 cubic feet of wood per acre annually. Nationally, for all forests, Site Class V includes 27 percent of the commercial forest but only 12 percent of the potential productive capacity. This capacity is a biological potential of the poorer sites; their economic potential is much smaller, since many costs are a function of land area rather than of output. Site Class V lands are not evenly distributed among the major forest ownership categories. If the Site Class V land is eliminated as not being in fact "commercial," the per acre potential increases by 20 percent for all forests, but the increase is 22 percent for national forests, 28 percent for other public forests, 11 percent for forest industry forests, and 19 percent for other private forests (table 1-5). If these less-productive forests are eliminated from the commercial classification, then much of the apparent disparity in average productive capacity among the major ownership classes disappears. It seems evident from these data that the forest industry firms sought to obtain more-productive forests, as would be expected, but, more significantly, to a large extent they avoided the least-productive forests. By comparison, the national and other public forests include many more Site Class V forests, which are not very productive for wood growing but are possibly of relatively high value for other forest outputs.

Table 1-5. Estimated Productive (Wood-growing) Capacity, Forests of Different Ownership Categories, for Site Classes I to V and I to IV, 1970

(cubic feet per acre annually)

Ownership category	Estimated productive capacity[a]	
	Land in Site Classes I to V	Land in Site Classes I to IV
National forests	76	93
Other public	72	92
Forest industry	88	98
Other private	74	88
All	76	91

Source: Report of the President's Advisory Panel on Timber and the Environment (Washington, D.C., 1972) p. 36.

[a] Based on capacity to grow industrial wood, in fully stocked natural stands, with good but not intensive management, at near the age of maximum mean annual increment. Capacity for ownership category calculated by multiplying midpoint of range (or estimated average of range, for open-ended classes) by acreage in each site class.

The picture for sawtimber (table 1-4) is generally similar to that of growing stock in overall trends, changes by periods, and relationship among major ownership categories. The national forests make a somewhat better showing for sawtimber than for growing stock, reflecting in part the management objective of the Forest Service to produce relatively large trees.

The harvest of hardwoods, both of growing stock and of sawtimber, in each period and for each major ownership category has been substantially less than net growth; and the disparity between growth and harvest has increased over the years. While there may well be some scarcity of high-quality hardwoods, the total hardwood supply is considerably above the demand for such hardwoods.

For softwoods, the picture is different and somewhat more complicated. In general, the harvest of softwood sawtimber in most periods and for most ownership categories has exceeded net growth by varying amounts. For growing stock, harvest exceeds growth in some periods and for some ownerships, and vice versa for other ownerships and other periods. In general, harvest of softwood sawtimber has declined relative to annual net growth in the more recent periods. For all periods, the harvest of softwood sawtimber from forest industry forests has exceeded annual net growth, and has been higher, relative to annual net growth, than for any other major ownership category. Table 1-2 showed that the inventory of growing stock per acre on forest industry forests has been relatively constant over the years. A constant inventory of growing stock and a harvest of sawtimber exceeding net growth are evidence of some continued shifting of forest industry forests toward younger-age stands.

WOOD GROWTH POTENTIAL

In any discussion of strategies for future forest production, the possible and probable annual growth of wood fiber must receive major consideration. Obviously wood is an extremely important output of forests; in many cases, its annual growth is more valuable than annual production of all other goods and services from the forest. Moreover, wood is typically the "paying partner"--the only or the chief enterprise that produces tangible income to the forest owner/manager.

Net Annual Growth

Consideration of future potential may well start with information on the present. The net growth of potentially usable industrial wood in 1977 was 21.7 billion cubic feet (USDA, Forest Service, 1980). In order to understand the meaning of this figure, it is necessary to consider carefully what is included, and what is excluded. The figure represents the amount of wood grown in 1977 on what the Forest Service defines as growing stock: that is, trees of commonly used industrial species; of sawtimber and poletimber sizes (meaning trees 5 inches or more in diameter at breast height -- 4 1/2 feet above ground); live trees, excluding cull trees; volume measured up to a 4-inch top or to the point where the trunk breaks into limbs; on land capable of producing 25 or more cubic feet of industrial wood annually in a fully stocked natural stand at about the age of mean annual increment of growth; growth based on present forest type, even though some other type might produce more wood annually; and net growth, above mortality from any cause except harvest. This definition includes no consideration of the profitability of either harvesting a present timber stand or of growing another stand on the same site, and net growth may be either more or less than harvest in any year.

The definition of net annual growth given above excludes much wood on all forest sites: the wood below a normal stump (1 foot above ground) in both stem and roots; the wood in the main trunk and the limbs above the point where the trunk breaks into limbs; dead trees, cull trees, and deformed trees that could not be used for industrial wood purposes; saplings, or trees more than 1 but fewer than 5 inches in diameter; seedlings, or trees less than 1 inch in diameter; and trees of nonindustrial wood, which in practice means trees of species used for fuel. Reserved forest lands, such as national parks and wilderness areas established by congressional action, are excluded from the commercial forest category and from these data.

A review of the wood excluded from the net annual growth definition suggests that annual wood growth in the total tree biomass is substantially larger than the net growth, by a factor of probably 2 or even more, on the average, although this of course varies considerably from one forest tract to another. The Forest Service reported a total mortality in all forests and from all causes of 3.9 billion cubic feet of wood in 1976 from trees of growing stock size. Gross growth of trees of these sizes and species and of the parts of trees listed above was thus 25.6 billion cubic feet (USDA, Forest Service, 1980).

Management Practices Affecting Potential Growth

A crucial consideration in future timber growth is the ability and willingness of forest owners/managers to invest management, labor, and capital in the growing of timber (Clawson, 1977, especially pp. 35-37). Although timber harvest may either exceed or fall short of net annual growth in the short run as timber inventories are drawn down or built up, in the long run harvest and growth must average out equally. Some regrowth of trees will occur on normally forested sites in spite of neglect by an owner/manager and sometimes in spite of efforts to prevent tree regrowth. That regrowth may fail to take full advantage of the productive capacity of the site. Beyond this "naturally" occurring tree growth, more wood growth can be produced by various management practices.

Forest owners/managers willing to grow timber for future harvest may adopt any or all of various practices to achieve that goal. First, and probably most important quantitatively in the past, is the institution of fire control by forest owners as relatively large groups, perhaps operating through state or other government. Forest fires, being notoriously disrespectful of property lines, must be fought without respect to such boundaries, and hence the need for group action in fire control. Individually, forest owners/managers can decide to refrain from setting fires on their own property. (At one time uncontrolled burning of woods was very common, especially in the South, although it is much less so today.) In choosing not to burn, the forest owner/manager surely has some concept of future wood production in mind. A somewhat similar situation arises on some farms with woodlots; exclusion of livestock during the early stages of seedling growth, which also represents a management decision, can make a significant difference in the amount and species of the timber regeneration, and hence of future wood production. The planting of seedlings, or the less common but somewhat similar planting of seeds, is another major management practice full of implications for future timber growth. (Although it receives much attention in professional and popular literature, this measure is only one of many, as this listing indicates.) The owner/manager may also decide to reduce greatly the interval between harvest of an old stand and establishment of a new stand, thus adding to average net annual wood growth. Thinning or selective harvest of trees of some sizes or some species or both is another practice that affects growth, especially for the longer run. (By careful periodic thinnings, most of tree mortality can be captured for usable wood supply--trees are removed before they die, and thus net wood growth approaches or equals gross wood growth.) Lastly, regarding when to harvest timber, the management decision which includes not harvesting until the trees have reached something approximating economic maturity is of importance for future timber supply, since premature harvest results in reduced average annual growth per acre.

Any decision to grow more timber, either by letting "natural" forces operate unhindered or by using measures designed to increase growth, will take time to show results, and the length of time will, of course, vary from practice to practice. Those who discuss forestry wholly in terms of tree planting emphasize the long periods involved in timber production. Much literature focuses on planting seedlings and on the time needed for the tree to mature, which may range from perhaps 20 years for some species grown for pulpwood to 100 or more years for other species grown for sawtimber under extensive forest management practices. But there are other practices that can increase timber output within a much shorter time. For instance, thinning some stands may produce a significant immediate harvest at no reduction in future net wood growth. On the other hand, the development of genetically superior tree varieties (within any species) may take a few generations, although even here certain management practices can shorten the time required. Rather than generalizing about time requirements or increased timber production, it might be well to try to measure actual responses over varying time periods.

There are, in general, two possible approaches to estimating the willingness of forest owners/managers to grow timber for some future harvest. First, one may study the past, particularly the practices forest owners/managers used in their timber production, and then relate this to the prices they expected or received for harvested timber. This approach has the great strength of being based on actual forest owner/manager actions, but it also has the considerable weakness of trying to sort out many past events and to separate the influence of one factor such as price from that of other factors. Furthermore, even if one can closely estimate past events, this is no proof of what the future will be. A second approach is to build engineering-silvicultural-economic models to estimate what returns a forest owner/manager might reasonably expect at some defined date in the future from some present forest management action. This approach has great value in looking ahead, not backward, and in being able to separate the influence of different management actions, but it has a very serious weakness --forest owners/managers may not respond to price or other stimuli as an analyst predicts. Let us consider the second approach, then the first.

Engineering-Silvicultural-Economic Models

A committee of the National Academy of Sciences/National Research Council has explored the biological wood-growing potential of U. S. forest, as tempered somewhat by economic and management considerations (CORRIM, 1976). The Committee on Renewable Resources for Industrial Materials (CORRIM) considered various intensive forest management practices and also both economic and management constraints on the use of such intensive practices, and concluded that "the biological productivity of the commercial forest lands of the U.S. is such that, where economic and social conditions permit, the net realizable growth of these forests could be doubled within half a century by the immediate widespread application of proven silvicultural practices" (CORRIM, 1976, p. 7). CORRIM utilized proven forest management techniques in estimating wood yield responses to various management practices, but also made allowance for the fact that such practices would not be applied, for various reasons, to all forest land on which they would be profitable. Combining the various increases in wood growth due to more intensive management with their caution about the economics of such practices and the management problems of getting them applied everyhwere, CORRIM's final estimate of annual net growth is more than 30 billion cubic feet for all species.

The Forest Industries Council (1980) has estimated the additional wood that could be grown on forests of different ownerships by application of various forest management practices, subject to the constraint that all such practices should yield a 10 percent or higher return on the additional investments and expenditures necessary to carry them out. The report includes no explicit time frame for the achievement of the results, nor is there any estimate of the amount of additional wood that could be grown for a return greater than or less than the 10 percent standard. These estimates were made for the twenty-five most important forest states, which contain about 83 percent of all commercial forests in the country. The result is that 10.9 billion cubic feet of additional wood growth is obtainable under the conditions specified. The sources of this added supply are highly significant: 4.5 billion cubic feet come from early harvest and prompt regeneration: 3.3 billion cubic feet from stand conversion; 1.8 billion cubic feet from stand regeneration; 0.9 billion cubic feet from intermediate stand management; and 0.3 billion cubic feet from planting idle cropland (Forest Industries Council, 1980, p. 46). The discussion does not make clear exactly what role thinning and fertilization are expected to play, and there seems to be no explicit consideration of genetics, including seed selection from superior trees. Although an explicit time frame is not given, it appears that considerable increases in net annual growth could be achieved in a comparatively few years.

Perhaps the most significant aspect of this study, in addition to its estimates of added wood supply by management practice and by ownership groups, was that it involved such a large number of foresters, especially from forest industry firms but also from public forestry agencies, on a state-by-state basis. The results represent the consensus of a very large group of the best informed foresters in the country.

Hyde (1980) has explored in detail the management, the ecological, and the economic aspects of many forms of intensive forest management in the Douglas fir region. The length of time between timber harvest and the reestablishment of a productive and thrifty stand on a particular site is very much a matter of management policy and of coordinated operation of various parts of the forest management enterprise. The number of seedlings planted per acre is variable and is also a means of intensification of forest management. Thinning of growing stands is a means of capturing for economic return the trees that would otherwise be part of the normal mortality; in this way, net wood growth can be made to more nearly equal gross growth. Fertilization at intervals may increase tree growth. These and perhaps other practices can be combined in various packages; in general, in forestry as in farm crop production, each management practice reinforces the other, and the package is often more productive than the sum of the individual practices.

Hyde also considered the potentials of genetic improvement of the trees planted. His analysis is particularly strong in its consideration of the costs and returns from various practices on sites of different inherent productive capacity. The effect of the package of intensive practices is not only to grow much more wood per acre annually over the rotation cycle, but also to shorten that cycle by many years. At the same time, thinning and other practices result in trees as large at the end of the shortened intensive cycle as would otherwise result after many more years under natural forestry. The intensive forestry thus outlined is more costly than at least some of the natural forestry, but may be economic at expected future prices for wood. At the same time that much more wood is grown by intensive forestry on the more productive forest sites, the less productive sites are excluded from timber management because the prospective returns do not warrant the estimated costs. The net result of Hyde's analysis is a potential (generally economic) supply of 4.4 billion cubic feet annual wood growth (Hyde, 1980), contrasted with a net annual growth in the Douglas fir region (estimated by the Forest Service) of 1.9 billion cubic feet in 1976.

The foregoing are examples of the engineering-silvicultural-economic approach described earlier. The weakness of this approach, as was indicated, is that the estimates of rational behavior implicit in these models may not include all the factors, or may weigh factors differently, than forest owners/managers will do in practice. This weakness therefore leads to an examination of past trends in wood growth in relation to past prices of stumpage.

Examining the Past

In the South, the annual net growth of softwoods (mostly pine) on private nonindustrial forests increased greatly between 1952 and 1977 while at the same time prices of pine stumpage were rising. If net growth in each year for which data are available is related to average prices approximately 10 years earlier, a moderately close relationship results (figure 1-9). If net annual growth is related to average prices approximately twenty years earlier, the relationship is even closer and the output at any price level is substantially greater. It will be recalled that this class of owners has about 71 percent of all forests in the South; hence these relationships are important. No one questions that many factors in addition to price have affected net timber growth on these forests in this region over these years. But it is also highly probable that net timber growth would not have risen so much, if at all, if prices had not been generally favorable. While figure 1-9 shows what the overall experience has been in the past, it does not show the relative contribution of different forestry

Price of Southern
pine stumpage -
1967 dollars
per 1,000 bd. ft.

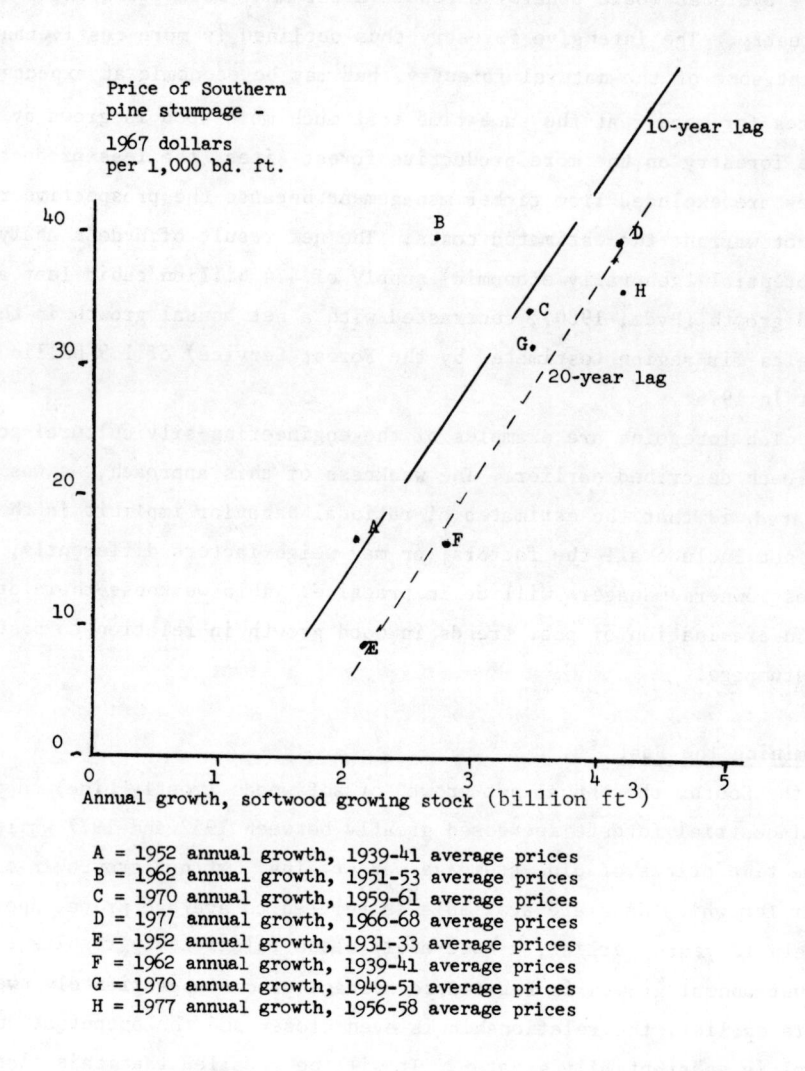

A = 1952 annual growth, 1939-41 average prices
B = 1962 annual growth, 1951-53 average prices
C = 1970 annual growth, 1959-61 average prices
D = 1977 annual growth, 1966-68 average prices
E = 1952 annual growth, 1931-33 average prices
F = 1962 annual growth, 1939-41 average prices
G = 1970 annual growth, 1949-51 average prices
H = 1977 annual growth, 1956-58 average prices

Figure 1-9. Price-growth relations for Southern pine, 1952 to 1977, nonindustrial private forests

Source: Marion Clawson, The Economics of U.S. Nonindustrial Private Forests (Washington, D.C., Resources for the Future, 1979) p. 292.

practices to the net result. Moreover, even if these relationships existed in the past, that would not prove that they will continue in the future. However, if they do continue, stumpage prices actually received in 1978 would lead to an annual net growth of about 6.4 billion cubic feet of softwood in 1998, or an increase of 50 to 60 percent above present net growth.

A somewhat similar relationship exists for net growth of softwood growing stock and stumpage prices of Douglas fir on the Pacific Coast for the same group of forest owners and for the same period of years (figure 1-10). This class of forest owners is less important on the Pacific Coast than in the South. The relationship between net growth and prices ten years earlier is about the same as with the prices twenty years earlier, except that in the latter case a much greater output is realized at a given price level. Prices of Douglas fir stumpage have already risen sharply along the Pacific Coast in recent years, and if past growth-price relationships hold in the future, by 1998 these forests will be also producing about 50 to 60 percent above their present level.

Figures 1-9 and 1-10 are based on net growth of timber. The growth might not all be harvested; some might have been added to inventory. And of course much timber harvest might have come out of inventory, so that these data on net growth are not exactly indicative of harvest in the short run. However, the net growth is presumably responsive to the forest owner/manager's decision to invest in further timber growing.

Constraints on Estimates of Potential Growth

All of the estimates of potential future net timber growth presented in this section have some limitations--of time for the results to be achieved, of economic results that must be attained, of practical management limitations, or others. If one sought to estimate the ultimate future growth of wood fiber annually under no limiting assumptions as to tree species or genetic inheritance, and with all management practices subject to innovation and change, and without concern for costs in relation to returns, then the estimate of ultimate production potential would have to be at least double, perhaps much more, any of the foregoing estimates of "potential." That is to say, "potential" is a relative term, and potential as estimated under one set of assumptions may be substantially less than potential under other assumptions.

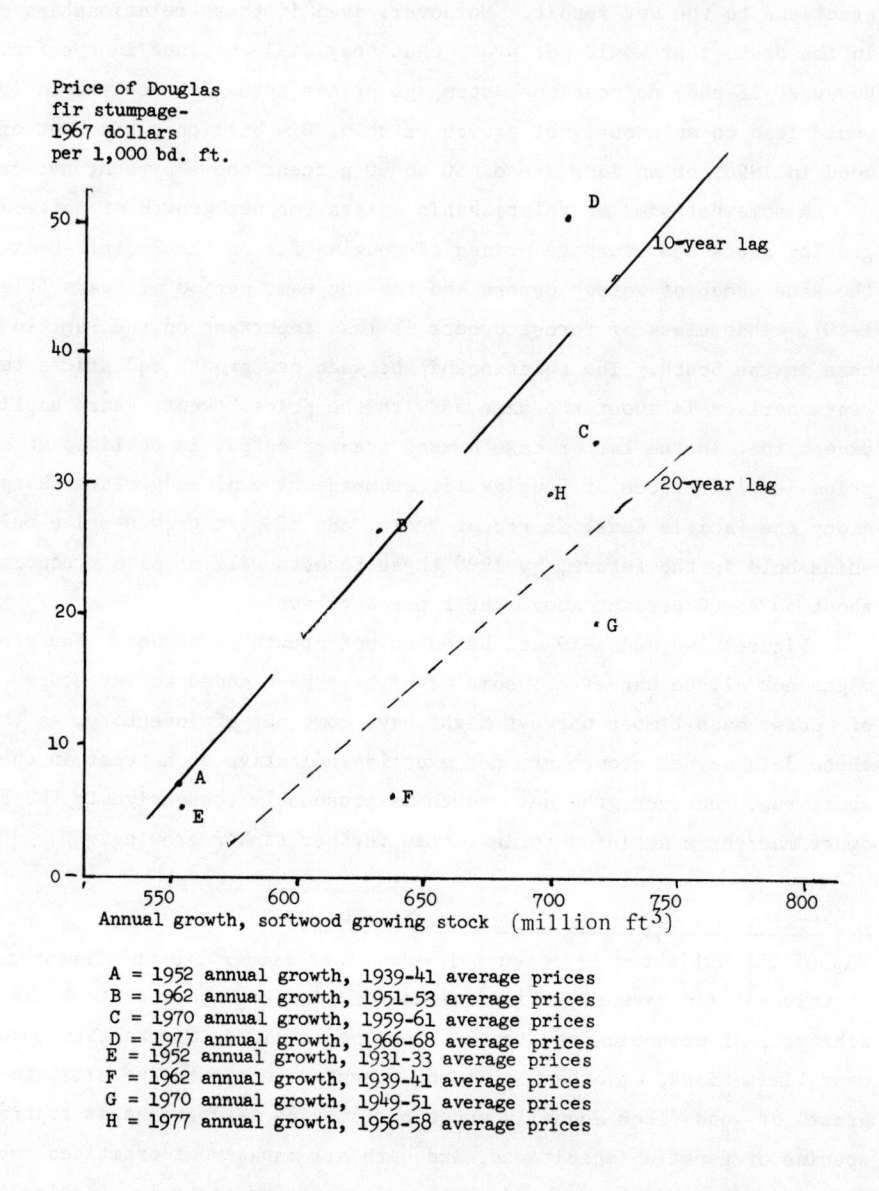

Price of Douglas
fir stumpage-
1967 dollars
per 1,000 bd. ft.

A = 1952 annual growth, 1939-41 average prices
B = 1962 annual growth, 1951-53 average prices
C = 1970 annual growth, 1959-61 average prices
D = 1977 annual growth, 1966-68 average prices
E = 1952 annual growth, 1931-33 average prices
F = 1962 annual growth, 1939-41 average prices
G = 1970 annual growth, 1949-51 average prices
H = 1977 annual growth, 1956-58 average prices

Figure 1-10. Price-growth relations for Pacific Coast, 1952 to 1977, nonindustrial private forests

Source: Marion Clawson, The Economics of U.S. Nonindustrial Private Forests (Washington, D.C., Resources for the Future, 1979) p. 293.

The material presented in this section is summarized here and in table 1-6. The CORRIM report envisaged a considerable potential increase in net annual growth of wood; at the time that report was prepared, the latest available figures on net timber growth were for 1972, at which time growth was 19.8 billion cubic feet annually. The CORRIM estimate thus was for an approximately 50 percent increase in potential output "within half a century." The Forest Industries Council figure is for about the same percentage increase; although the report is not explicit as to time, presumably the increase could be achieved in a shorter time than CORRIM specified. Moreover, the Forest Industries Council figure is higher (or more optimistic) in at least two other respects: (1) it does not include the small but possibly significant increases that could occur in the twenty-five less important forest states that were not included in its analysis, and (2) it is subject to a rather rigorous economic test of a 10 percent or higher interest return on capital investment and expenditures. Hyde's estimate for the Pacific Northwest, not precise as to timing, is for a somewhat larger percentage increase. My estimates of probable production in 1998 from private nonindustrial forests in the South and in the Pacific Northwest, based on continuance of past growth-price relationships, are for increases somewhat in excess of 50 percent. To this extent, these various estimates of "potential" net growth of wood a couple of decades or so in the future are in rough agreement, in spite of differences in methodology, areas, and ownership categories considered, and perhaps other factors. Given the unavoidable uncertainties about the future, the degree of agreement is considerable.

As has been mentioned, each of these estimates of "potential" has time, economic, or other constraints. If one tries to remove all such constraints and to visualize a future in which all aspects of forestry are subject to change, a very much larger total net wood annual growth is possible, as the last line of table 1-6 illustrates. I find it noteworthy that trained foresters in the past have repeatedly underestimated future timber growth, including biological potential wood growth (Clawson, 1979b). Imagining a distant future is always difficult and subject to large margins of error, but estimates being too low is at least as likely as their being too high.

Table 1-6. Annual Present and Alternative Future Growth of Potentially Usable Industrial Wood and of Total Tree Biomass in the United States

(billion cubic feet)

Item	Annual net growth of potentially usable industrial wood	Annual growth, total tree biomass
1. Present (1977)	21.7	45 +
2. CORRIM estimate of biological productive potential	30 +	--
3. Forest Industries Council estimate of potential increment to growth, 10 percent of greater return on investment	10.9	--
4. Hyde estimate of additional growth in Pacific Northwest	2.5	--
5. Estimated growth in 1998 on nonindustrial private forests, if past growth/price relations continue:		
South	6.4	--
Pacific Coast	1.3	--
6. Ultimate potential annual growth, with genetic improvments, unlimited intensive management practices, and optimum timber rotations	60 - 100	100 - 200

Sources: 1. (col. 1) U.S. Department of Agriculture, Forest Service, An Analysis of the Timber Situation in the United States, 1952-2030, Review Draft (Washington, D.C., 1980); (col. 2) Clawson estimates. 2. Committee on Renewable Resources for Industrial Materials (CORRIM), Commission on Natural Resources, Renewable Resources for Industrial Materials (Washington, D.C., National Academy of Sciences, 1976). 3. Forest Industries Council, Forest Productivity Report (Washington, D.C., National Forest Products Association, 1980). 4. William F. Hyde, Timber Supply, Land Allocation, and Economic Efficiency (Baltimore, Johns Hopkins University Press for Resources for the Future, 1980). 5. and 6. Clawson estimates.

Note: Dash = not applicable.

Capacity and Willingness to Grow Wood: More Considerations

All of the material so far in this section has dealt with capacity and willingness to grow wood. It is frequently objected that much of the wood likely to be grown will not be harvested--that forest owners/managers will be unwilling to harvest their timber. Certainly the management of the national forests has been based on an unwillingness to harvest timber faster than a rate that can be sustained in the future--not only sustained yield but also even flow--and generally at a rather extensive level of forest management. It is generally assumed that forest industry firms will harvest all or nearly all the timber they grow, on schedules consistent with their maximizing economic return from woods, mills, and marketing organizations. The real question about willingness to harvest thus arises for the nonindustrial private forest owners.

A number of surveys of private nonindustrial forest owners have revealed considerable percentages of owners who say they are opposed to timber harvest on their forests. Repeat surveys of owners, in the relatively few cases in which they have been made, show that many owners who had said they would not harvest timber have, in fact, done so. Surveys of owners' intentions may be more misleading than helpful. The basic difficulty is not that people change their minds or are dishonest in reporting, but that timber harvest from typical nonindustrial forests occurs at intervals of some years during which ownership of the forest, income and tax situations of owners, and economic factors generally may change. "In a great many forest situations in the United States, trees will be cut for sawtimber when 40 to 100 years of age; sales from smaller forest properties may occur at intervals of 11 to 33 years; and average tenure may be only 7 years" (Clawson, 1979a, p. 257).

In order for a nonindustrial private forest owner to sell timber, a minimum of two conditions must be met: (1) the owner must have a volume and quality of timber that, considering the general market, the particular location of the owner's forest, and perhaps other factors, are attractive to potential timber buyers and processors; and (2) there must be a buyer. These conditions are not present for many forest owners in many locations at various times. Unless they are met, it is idle to speak of the forest owner's willingness to sell timber.

My conclusion is that all truly merchantable timber (the term deals with location as well as volume and quality) will ultimately be sold, by some future owner of the forest if not by the present one. The timing of such timber sales may be different from the timing a silviculturalist or economist would have chosen, but, averaged over a large number of owners of such forests and over a considerable period of time, the results may not be much different from what they would be if either silvicultural or economic considerations had governed timber sales.

A consideration of lumber production, new housing starts, and lumber prices in one relatively recent housing construction cycle suggests that lumber supply is highly elastic when capacity to increase lumber production exists (figure 1-11). The data in this chart suggest that it is not so much higher prices that draw out larger lumber supply, as it is the ability to sell more lumber at the same or possibly slightly higher prices. In some of the other building cycles since 1950, this high elasticity of lumber supply is less evident. In some situations, where there is no idle or excess capacity to increase lumber production, price rises have been very great while supply response has been limited. Figure 2-11 measures lumber output, which is a response of lumber mills; but the lumber is made from logs, which may have originated from the mill owner's forest or may have been bought from some other forest owner, public or private. In either case, there must have existed a considerable willingness to harvest logs from standing timber, within the limits of the variable output shown in figure 1-11.

FOREST OUTPUTS OTHER THAN WOOD

It is both commonplace and accurate to assert that forests provide many valuable goods and services in addition to wood--water, wildlife, wilderness, recreation, scenic values, and others. But it is always impossible to treat these other goods and services with the same detailed analysis as can be given to wood, for the simple reason that the necessary data are lacking. Data on forest land area, timber stand, wood growth, wood utilization, wood prices, and other aspects of wood production from forests are always less in volume and poorer in quality than is ideal; but compared with data on the nonwood outputs of forests, the data on wood are very

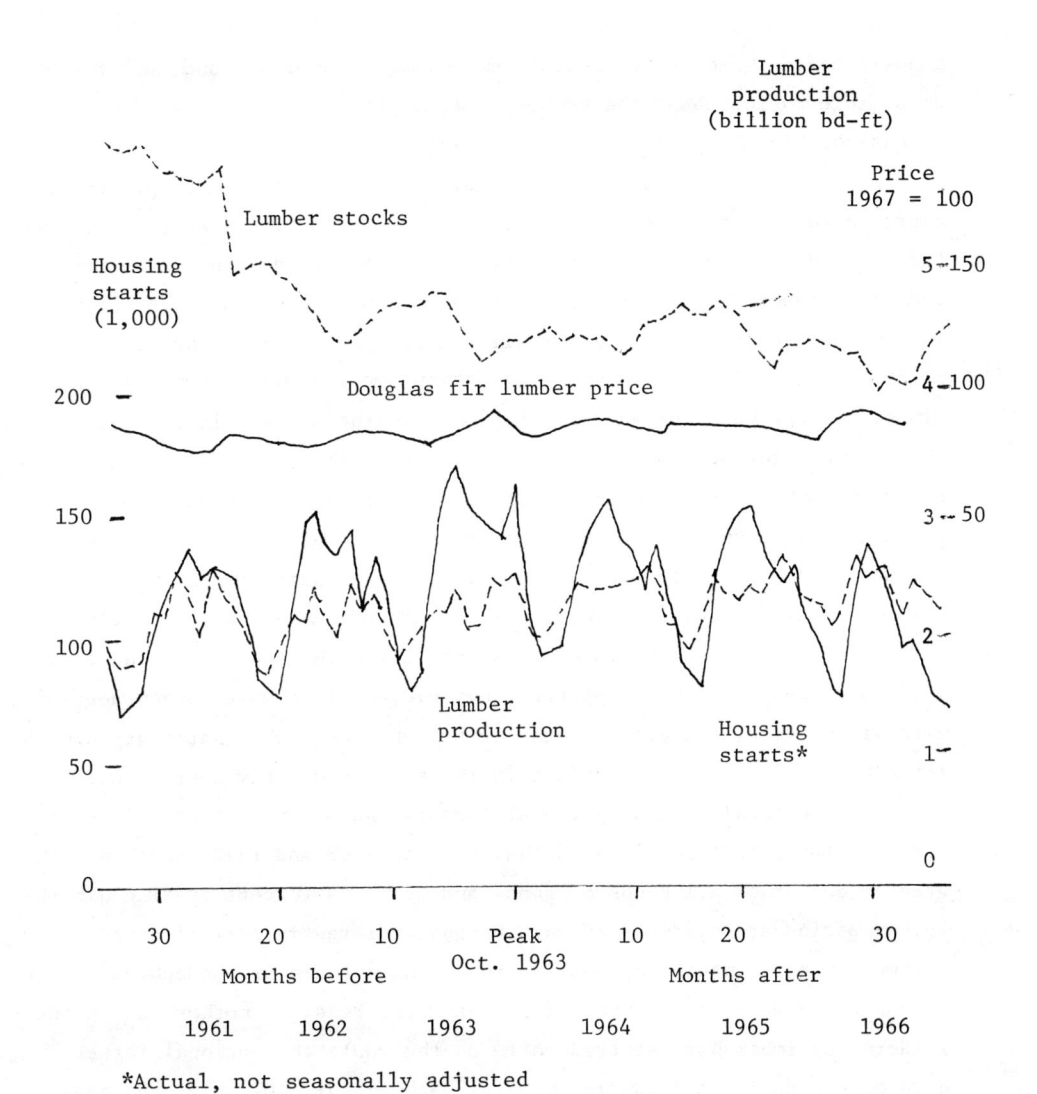

Figure 1-11. Actual housing starts, softwood lumber production, softwood lumber stocks, and price of Douglas fir lumber, Housing Cycle #5, November 1960 to November 1966

Source: Marion Clawson, unpublished RFF manuscript.

good indeed. There is, of course, much common knowledge and much anecdotal information about the nonwood outputs, but comprehensive data for a considerable period of time are totally lacking.

On the basis of available information, it is possible to present an approximate picture of the output of the national forests at two different time periods (table 1-7). Apparently the greatest increase in output has been in outdoor recreation; specific data on wilderness use are lacking for the earlier date, but use of wilderness areas probably has increased relatively as much or more than have other forms of outdoor recreation. Timber harvest has increased greatly because the harvest in the 1920s from the national forests was at a very low level. The best indication is that wildlife numbers on national forests have also increased; data not shown in table 1-7 suggest that the amount of forage consumed by game on national forests increased sevenfold in approximately the same time span. The volume of water flowing off national forests probably changed little, but the use of this water increased substantially. These data are the best available, and the general picture they present is almost surely accurate, even if the details are fuzzy--the output and the use of national forests for almost all goods and services increased considerably over a forty-year period. The total area of national forests changed very little from the period between 1925 and 1929 to that between 1968 and 1972, and hence the greater output of all kinds of goods and services (except grazing) in the <u>latter</u> period are evidence of more intensive forestry for all outputs and not merely more intensive wood growing. While there are undoubtedly situations where more of one kind of output means less of another, as in the wilderness/timber harvest trade-off, on the whole the national forests have been capable of producing more of each output. Clearly, national forest management is not a zero-sum game in which winnings by one group come only at the expense of losses by some other group. Everyone can "win" if sufficient intelligence and goodwill are invested.

Data are lacking to provide a comparable analysis for any other major forest ownership category. State forests to some extent and state parks to a great extent have experienced a great increase in recreational use. The same is true of federal forests other than national forests. There is reason to believe that this is also true of forest industry forests.

Table 1-7. Average Annual Harvest of National Forest Resources, 1925-29 and 1968-72

Resource	Annual average 1925-29	Annual average 1968-72	1968-72 as multiple of 1925-29
Timber cut (billion bd-ft)	1.35	11.54	8.6
Recreation visits (millions)	6.3	188[a]	30
Wildlife (thousand big game killed by hunters)	216[b]	582	2.7
Water	--	--	Probably 2.0[c]
Forage for domestic livestock (million animal unit months)	12.67[d]	8.60[e]	0.68

Source: Marion Clawson, The Economics of National Forest Management (Washington, D.C., Resources for the Future, 1976) p. 20.

Note: Dash equals not available.

[a] Million visitor days, 1973.

[b] 1940-44 average.

[c] There are no data available on use of water flowing off national forests. With the volume of dam building, public and private, use in the latter period can hardly be less than double the former, even if total stream flow is unchanged.

[d] Figure is for 1929.

[e] Estimated from data on livestock numbers in Agricultural Statistics, 1974, p. 552, times average length of grazing season in 1964, as calculated from data in Clawson, 1967, p. 580.

There has been greater use of the water flowing from all forests. It is generally believed that the "other private" forests provide a considerable amount of recreation for their owners and their guests, as well as some for trespassers, but specific data are lacking. Ownership of forest land has provided both a protection from inflation and the source of substantial increases in real wealth--gains that may well exceed the value of the wood produced on other private forests.

In conclusion, however, we cannot be more specific and more quantitative about the goods and services, other than wood, produced by forests of any major ownership other than the national forests.

STRATEGIES FOR THE FUTURE

I conclude with several general statements, which are to some extent a summary of earlier parts of this chapter, to some extent projections of past trends, to some extent forecasts of probable future developments, and to a degree some (more or less gratuitous) advice as to a desirable national policy for our forests.

    1. The forests of the United States currently produce many goods and services--wood of various kinds for various uses, water, wildlife, wilderness experience, recreation, scenic values, and others.

    2. The output of each kind of goods and services has been rising in either volume or value or both for many years.

    3. The output of every good and service is far below its potential by varying amounts for different goods and services and by varying amounts in different locations.

    4. The concept of "potential" is not simple, and in fact our ideas of forest potential have undergone considerable change in the past and are likely to change further in the future.

    5. Economic and other forces of the recent past and of the present will lead to substantial further increases in forest output, of most or all kinds of goods and services.

    6. The demand for industrial wood (for lumber, plywood, and pulp) will continue to rise in the United States as population increases, per capita incomes rise, and environmental concerns provide an impetus for shifting toward renewable rather than to nonrenewable resources; and the growth and harvest of industrial wood both can and will increase also. In

keeping with trends of the past, the increased output will come largely from more intensive forestry on the more productive forests; there will be less or probably no increase in output from the less productive forest sites.

7. The demand for wood for fuel will surely rise just as it has risen substantially in the past few years. Two quite different situations should be distinguished: the harvest of fuelwood from existing forests, and the establishment of new fuel forests. The increasing volume of unharvested hardwoods, often of rather low quality for timber manufacture, offers a considerable opportunity for more wood for fuel without much affecting the supply of usable industrial wood. New fuel forests will require land, capital, labor, and management with the distinct possibility of impacts on the output of other goods and services competing for the same productive factors.

8. If the output of nonwood goods and services from the forests of the United States is to keep up with the growing demand, at least two factors need to change: first, there should be a better identification of the best forest sites for these other outputs and a better development of the trade-offs among them and between them and wood harvest for each site class; and second, in some way the owners/managers of forests must be able to realize more of the economic values from these nonwood outputs than they now receive. Unless, or until, the latter is accomplished, nonwood outputs will be forthcoming sparsely and reluctantly from both public and private forests. As it is now, those who gain these nonwood values assert the value of such goods and services but pay little directly to encourage forest owners/managers to produce such outputs.

9. For the publicly owned forests, the big challenge of the next decade or two is the full implementation of the National Forest Management Act of 1976 and similar legislation applying to other forests to bring about a greater degree of economic management. If the management programs are truly economically sound, and if some means can be devised to provide the managers of such forests with a return from nonwood outputs, then I have faith that the necessary appropriations for the management of public forests will be forthcoming.

10. For the nonindustrial private forests, the best public action is to leave them alone, aside from fire control. Increased and more efficient markets for output of such forests would greatly stimulate better management.

REFERENCES

Agricultural Statistics, 1974. 1974 (Washington, D.C., Government Printing Office).

Clawson, Marion. 1967. The Federal Lands Since 1956: Recent Trends in Use and Management (Baltimore, Johns Hopkins University Press for Resources for the Future).

---. 1976. The Economics of National Forest Management. Working paper EN-6 (Washington, D.C., Resources for the Future, June).

---. 1977. Decision Making in Timber Production, Harvest, and Marketing. Research Paper R-4 (Washington, D.C., Resources for the Future).

---. 1979a. The Economics of U.S. Nonindustrial Private Forests. Research Paper R-14 (Washington, D.C., Resources for the Future).

---. 1979b. "Forests in the Long Sweep of American History," Science vol. 204 (June 15) pp. 1168-1174.

---. 1981. "Competitive Land Use in American Forestry and Agriculture," Journal of Forest History vol. 25, no. 4, pp. 222-227.

---, R. Burnell Held, and Charles H. Stoddard. 1960. Land for the Future (Baltimore, Johns Hopkins University Press for Resources for the Future).

Committee on Renewable Resources for Industrial Materials, Board on Agricultural and Renewable Resources, Commission on Natural Resources. 1976. Renewable Resources for Industrial Materials (Washington, D.C. National Research Council/National Academy of Sciences).

Forest Industries Council. 1980. Forest Productivity Report (Washington, D.C., National Forest Products Association).

Hyde, William F. 1980. Timber Supply, Land Allocation, and Economic Efficiency (Baltimore, Johns Hopkins University Press for Resources for the Future).

Report of the President's Advisory Panel on Timber and the Envrionment. 1973. (Washington, D.C.).

U.S. Department of Agriculture, Forest Service. 1980. An Analysis of the Timber Situation in the United States, 1952-2030. Review draft (Washington, D.C.).

Bruce R. Lippke

COMMENTS

My comments will focus on the critical importance of having data and understanding that elucidate, insofar as possible, the long-term issues in forestry as we make plans and decisions in the present. Whether or not we have adequate data for long-term understanding, decisions and plans affecting the long term are being made--that is the nature of forestry. We are harvesting today the investment (or lack thereof) made decades ago. And likewise, what we do now will have impact not only in the short term, but decades hence.

As an industry economist, I am well aware of the almost exclusive concentration at the local level on the here-and-now issues relating to forestry, although perhaps the new-found interest in forest land use policy by local public bodies will better focus their attention on long-term matters. It has been common for forest economists, when called upon to testify before such groups, to be asked questions relating to immediate issues--such as this year's revenue estimate or trade flow and employment outlook. But they are not usually questioned, for example, on the long-term investment that must be put in place to support those revenues or jobs.

If we could understand the value structure behind delivering timber into world markets and the value system of alternative uses, we could possibly provide some rationale for better forest land use policy with highest economic use allocation. But if we are to succeed, the perspective of that value structure has to include a very long term understanding of world demand and supply.

In the 1930s, when U.S. farm output had grown very slowly for the preceding fifty years (not much faster than acreage cultivated--and that means no yield gains), perhaps no one would have forecast a doubling of output by 1980 with no further increase in acreage. And perhaps no one should have cared, because farm investments do not need fifty years to achieve full production. They produce in a couple of years. That, however, is a luxury we clearly lack in forest land management. Today we are harvesting the investment of long years ago with inadequate data on the nature of that investment or its productivity. Only recently do we have surveys on what is being invested now and the technology for understanding how to change our productivity. But again, regardless of whether or not we have adequate data for long-term understanding, decisions and plans affecting that long run are being made.

Marion Clawson's contribution to this volume properly takes the long-term perspective. He has characterized the productive potential of U.S. forest land by relying heavily on an evaluation of past changes in growth, changes we have already experienced. And Roger Sedjo in the next chapter adds a world context to the discussion and characterizes the production potential of some of the world forest lands with a stronger emphasis on the application of technology to plantations. Both are critically important in developing an understanding of the future when such long-term planning horizons are involved.

The U.S. forest products industry leaders, through the National Forest Products Association, have established a goal of developing trade surpluses of forest products, which involves higher production and is consistent with a goal of the lowest practical consumer cost. From my perspective as an industry economist, I can say that there is no question about our knowing how to produce more technologically. The technology is here to grow more wood and to produce more products from each unit of volume. The doubter need only go on tour to see the various applications of intensive management of the forest lands, and higher-technology mills and woods operations.

But beyond the question of the capability to produce more wood, which we can answer affirmatively, lie other serious questions whose answers are not clear: Are there the markets to absorb what could be grown? Is there enough confidence in these markets to support the economics of higher-production strategies? (Understanding markets and knowing which region is the cheapest supplier for each market are critical missing links in the research that is required to support better forest land use planning.) How does the potential to produce much more in the future fit with the other supply regions in servicing world markets? Will the values and the markets be there many years in the future to motivate economic investment in technology today?

In chapter 1 there are references to the Forest Service's market analysis. The timber market model used in the Forest Service's assessment includes base case results with the United States importing more and more wood over time. If we can produce much more technologically, we could hardly say that we are more dependent upon international suppliers. But the Forest Service base case results do assume that we are not competitive enough to keep imports from growing. However, the Forest Service also demonstrated with their model what should happen if more intensive forest management is practiced as an economic response to price increases. When rising prices were allowed to trigger investments in more intensive management with adequate rate of return, the results were totally different. Growth and production were much higher, the Canadian imports to the United States were forced out, implying a shift toward a trade surplus, and the simulated prices decreased as the volume response increased.

The growth assumptions used by the Forest Service for the simulation were still conservative compared with assumptions identified in a Washington state timber productivity study as current management levels. They were also conservative compared with management levels identified in an Oregon State University study as the low end of that study's assumed range for management intensification, which was based on a survey of land owner practices.

So the range of possible futures is enormous. These widely different views of the future use of our forest lands may suggest that we are not coping very well with the pressures on forest land, but they do illustrate critical elements of the information base necessary to support better public and private strategies for the future use of forest lands. Much has been done, but much is left for future research and discussion. We can produce much more of a renewable solar resource and substitute for other energy sources, but will we?

Chapter 2

THE POTENTIAL OF U.S. FOREST LANDS IN THE WORLD CONTEXT

Roger A. Sedjo

To understand the pressures on U.S. forest lands, it is necessary to understand both the global context within which the United States is operating and certain fundamental changes that are occurring in the forest resource.

This chapter covers three main topics: first, it provides a quick survey of the global forest resource base and identifies major producing regions; second, it presents a broad conceptual framework for examining some of the fundamental changes that are occurring in the nature and mix of the world's forest resource; and third, it concludes with speculation as to the future U.S. role as an industrial forest resource producer within the context of a dynamic global environment. The focus of the chapter is limited to timber as a commodity, thereby abstracting from many important issues associated with environmental considerations and the noncommodity outputs of the forest.

THE RESOURCE BASE

The world's forest resources are widely distributed. All the continents share in holding the forest inventories (tables -1 and -2). However, the economic forest resources are found largely in the temperate regions of the Northern Hemisphere where the major industrial countries of the globe are located. North America, Europe, and the USSR account for more than 76 percent of total industrial roundwood production (table -3) and more than 88 percent of industrial conifer production in 1976 (table -4).

Table 2-1. Land Area of World Forest Resources, by Region and Type

Region	Coniferous[a] Land area (million ha)	Coniferous[a] Percent	Broadleaf[a] Land area (million ha)	Broadleaf[a] Percent	Combined coniferous and broadleaf[a],[b] land area (million ha)	Combined coniferous and broadleaf[a],[b] Percent
North America	400	35.2	230	14.1	630	22.3
Central America	20	1.8	40	2.4	60	2.1
South America	10	0.8	550	33.6	560	19.8
Africa	2	0.2	188	11.5	190	6.7
Europe	75	6.6	50	3.0	140	5.0
USSR	553	48.7	175	10.7	765	27.1
Asia	65	5.7	335	20.5	400	14.2
Oceania	11	1.0	69	4.2	80	2.8
Total world	1,136	100.0	1,637	100.0	2,825	100.0

Source: Reidar Persson, World Forest Resources: Review of the World's Forest Resources in the Early 1970s no. 17 (Stockholm, Royal College of Forestry, 1974).

[a]Closed forests.

[b]The totals for combined coniferous and broadleaf forests do not always add because no breakdowns have been given for areas in Europe and the USSR excluded by law from exploitation.

Table 2-2. World Growing Stock: Volume in Closed Forest

Region	Coniferous 100 million m³	Percent	Broadleaf 100 million m³	Percent	Combined coniferous and broadleaf[a] 100 million m³	Percent
North America	265	25.8	95	7.9	360	16.1
Central America	7	0.7	15	1.2	22	1.0
South America	5	0.5	595	49.6	600	26.8
Africa	1	0.1	51	4.3	52	2.3
Europe	80	7.8	40	3.3	120	5.4
USSR	612	59.5	120	10.0	733	32.7
Asia	55	5.3	285	23.7	340	15.2
Oceania	3	0.3	0	0	13	0.5
Total world	1,028	100.0	1,201	100.0	2,240	100.0

Source: Reidar Persson, World Forest Resources: Review of the World's Forest Resources in the Early 1970s no. 17 (Stockholm Royal College of Forestry, 1974).

[a]Sum of coniferous and broadleaf forests do not always add to total.

Table 2-3. World Industrial Roundwood Production, 1959 and 1976

	1959		1976	
Region	(1,000 m^3)	Percentage of total	(1,000 m^3)	Percentage of total
North America	350,108 (100.0)	35.7	455,643 (130.1)	34.0
United States	271,652 (100.0)	27.7	326,991 (120.4)	24.4
Canada	78,456 (100.0)	8.0	128,652 (164.0)	9.6
Europe (excluding Nordic countries)	125,940 (100.0)	12.8	176,168 (139.9)	13.1
Nordic countries[a]	68,225	7.0	88,125	6.6
USSR	267,000 (100.0)	27.4	302,932 (113.5)	22.6
Other	167,421 (100.0)	17.1	317,261	23.7
Total world	980,695	100.0	1,340,129	100.0

Source: Food and Agriculture Organization, Yearbook of Forest Products selected issues (Rome, FAO).

Note: Parenthetical figures give percentage change in output with 1959 = 100.

[a] Finland, Norway, and Sweden.

Table 2-4. World Production of Industrial Conifer Roundwood, 1959 and 1976

	1959		1976	
Region	(1,000 m^3)	Percentage of total	(1,000 m^3)	Percentage of total
North America	279,292 (100.0)	38.4	375,191 (134.3)	39.6
United States	205,830 (100.0)	28.3	225,729 (124.2)	27.0
Canada	73,462	10.1	119,462 (162.6)	12.6
Europe (excluding Nordic countries)	87,190	12.1	114,040	12.1
Nordic countries[a]	65,250	8.9	79,367	8.4
USSR	232,900	32.1	268,954 (115.5)	28.4
Other	62,088	8.5	108,193	11.5
Total world	726,720	100.0	945,745 (130.1)	100.0

Source: Food and Agriculture Organization, Yearbook of Forest Products selected issues (Rome, FAO)

Note: Parenthetical figures give percentage change in output with 1959 = 100.

[a] Finland, Norway, and Sweden.

Within the Northern Hemisphere, only a relatively few countries and even fewer regions are important forest product producers feeding world markets. The United States and Canada are important industrial wood producers, as are the Nordic countries and the USSR. About 63 percent of 1976 industrial wood production came from these four regions. The rest of Europe provided about 13 percent of the world's industrial wood production, while the East Indian archipelago of Southeast Asia supplied 5.2 percent of the world's industrial roundwood in 1976, or slightly more than the production of the Nordic countries. These regions together also account for the lion's share of forest products exports worldwide (Sedjo and Radcliffe, 1981).

Major world forest products trade flows move from the Pacific Northwest of the United States and British Columbia in Canada to the major consuming regions of Japan, the eastern United States, and Europe. The USSR provides forest resources to both the European and the Japanese markets. The Nordic countries feed forest products into the demanding European market. Finally, Southeast Asian tropical hardwoods flow into the U.S., Japanese, and European markets, sometimes directly as relatively unprocessed wood and also indirectly through processing facilities in East Asian countries. Besides the dominant producers and consumers, there are, of course, numerous other more modest supply regions and demanding countries.

Should the investigation be expanded to include nonindustrial wood--including, importantly, fuelwood--world production would be found to increase by about 90 percent and the traditional Northern Hemisphere producers would play a much smaller role in total production (table 2-5). However, the trade pattern would change little, since nonindustrial wood is rarely transported far enough to be involved in international transactions.

The United States Within the World Context

The United States stands as the world's leading producer of industrial roundwood and the second leading producer (behind the USSR) of all roundwood. Despite these high production levels, very high domestic demand levels have made the United States a net wood deficit region persistently importing more than it has exported throughout the century.

Historically, the United States has gradually shifted from being a wood surplus region to its current wood deficit status. Certainly the early settlers viewed wood as an inexhaustible resource, one that was for

Table 2-5. World Roundwood Production, 1959 and 1976

	1959		1976	
Region	(1,000 m^3)	Percentage of total	(1,000 m^3)	Percentage of total
North America	405,627 (100.0)	24.3	473,790 (116.8)	18.7
United States	319,421 (100.0)	19.1	341,397 (106.9)	13.5
Canada	86,206 (100.0)	5.2	132,393 (153.6)	5.2
Europe (excluding Nordic countries)	210,545 (100.0)	12.6	195,612 (92.9)	7.7
Nordic countries[a]	87,215 (100.0)	5.2	110,206 (126.4)	4.5
USSR	397,000 (100.0)	23.8	384,543 (96.9)	15.2
Other	569,728 (100.0)	34.1	1,360,077 (238.7)	53.9
Total world	1,670,115 (100.0)	100.0	2,524,219 (151.2)	100.0

Source: Food and Agriculture Organization, Yearbook of Forest Products selected issues (Rome, FAO).

Note: Parenthetical figures give percentage change in output with 1959 = 100.

[a] Finland, Norway, and Sweden.

the most part seen more as an obstacle to transportation and alternative land uses than as an asset. Given the almost ubiquitous feature of the forest resource and the relatively high costs of transportation of wood resources, it is not surprising that for much of our early history international considerations were relatively unimportant. Gradually but continuously, however, U.S. interactions with foreign forest product markets have been increasing. By the end of World War I the United States had become a net importer of pulp and paper products, and during World War II the United States became a net importer of lumber. That situation prevails to this day.

The long-term dependence of the United States on foreign timber resources is not so much a sign of the lack of domestic production as it is an indication of the high level of domestic demand and of the location of Canada with its vast inventories of old-growth forest along a major border with prime access to major U.S. markets. It should be noted that when North America (the United States and Canada) is viewed as a unit, it is the principal wood-producing and wood-exporting region of the world (Sedjo and Radcliffe, 1981).

Simultaneously, however, even while being a net wood importer, the United States still exports large amounts of wood and fiber products. In the early 1960s, the great Columbus Day storm swept the forests of the Pacific Northwest. The resulting salvage operation stimulated the then-fledgling log export market to Japan so that within a decade earnings from conifer log exports from the Pacific Northwest alone mounted to more than $750 million annually. In recent years the United States has also experienced earnings of over $1 billion annually from the exports of paper and paperboard products, while at the same time being a major importer of both wood pulp and newsprint. Today the United States is generally well integrated into world markets for forest products. Domestic prices of forest products can no longer be viewed as being wholly determined by domestic demand and supply, nor even by North American supply and demand, but must be viewed in a worldwide context.

Table 2-6 presents the U.S. forest products trade balance (in current dollars) for selected years between 1950 and 1976. For each of these years the U.S. balance was in deficit, and the balance exhibits considerable

Table 2-6. U.S. Forest Products Trade for Selected Years, 1950 to 1976

(millions of current and constant 1976 U.S. dollars)

Year	Exports	Imports	Net trade[a] (current dollars)	Net trade[a] (constant 1976 dollars)
1950	118	1,035	(917)	(2,234)
1956	321	1.460	(1,139)	(2,423)
1959	391	1,531	(1,140)	(2,260)
1967	986	1,938	(952)	(1,612)
1970	1,621	2,299	(678)	(994)
1971	1,481	2,683	(1,202)	(1,649)
1972	1,817	3,291	(1,474)	(1,975)
1973	2,663	4,016	(1,353)	(1,713)
1974	3,537	3,877	(340)	(391)
1975	3,505	3,951	(446)	(468)
1976	3,994	5,431	(1,437)	(1,437)

Source: Food and Agriculture Organization, Yearbook of Forest Products selected issues (Rome, FAO).

[a] Parenthetical numbers denote a deficit.

variability. The deficit (in current dollars unadjusted for inflation) shows a modest rise between 1950 and 1976. However, if the deficit is deflated by a common price deflator (in this case, the gross national product deflator), it exhibits a modest decline in real terms between 1950 and 1976.

Global Trends in Forestry

Within the global context, some important trends should be noted. First, the role of the traditional Northern Hemisphere temperate climate producer regions has been declining over time for both industrial and nonindustrial wood. In the nonindustrial area, the pressure of increasing population alone probably guarantees a greater utilization of wood for fuel and other nonindustrial purposes. The rising real prices of fossil fuels are certain to exacerbate this tendency. With regard to industrial wood, the developing regions of the world are putting increased pressure upon their own forest resources to feed the economic expansions that are occurring. In addition, where available, the forest resources of the developing regions are being increasingly utilized and exported to the major world markets.

These trends suggest a growing worldwide demand for both industrial and nonindustrial roundwood and an increased role that the nontraditional producers are likely to play in meeting the increasing demand, particularly the localized fuelwood demand.

Coincident with the shifting of the geographic base of the world's forest resource production from traditional temperate climate regions to regions in the developing world, a fundamental restructuring of the mix of forest types that supply the world's woodbasket is occurring. The most notable feature of this restructuring is an accelerated shift away from reliance upon natural forest and toward a dependence upon high-yielding plantation forests, many of which utilize high technological inputs to meet industrial wood requirements. These shifts are occurring only slowly in the context of the massive global resource, and the structural changes will become pronounced only over decades. Nevertheless, the transition phenomenon itself will have a profound impact upon our perception of the forest resource and upon a variety of facets related to shorter-term aspects of the forest resource. This transition and its implications, particularly for U.S. forest lands, are addressed in the following sections.

THE FOREST TRANSITION

Forestry today is experiencing a transition similar to that which occurred in agriculture much earlier in human history. Just as mankind has progressed from gathering and hunting to cropping and livestock raising, we are today experiencing a transition from having our wood needs met from "old-growth" naturally created forest inventories to a situation in which conscious decisions are made to plant, manage, and harvest a forest. Just as modern agriculture involves decisions as to location, crop type, technological inputs, and management mode, so too forestry decisions increasingly involve questions of forest plantation location, species choice, technological inputs, and management regime.

Much of this transition phenomenon can probably be best explained in terms of a simple stock-adjustment model (Binkley, 1979; Lyon, 1981) that can be applied either to a particular forest (given qualifications) or to the global forest. In terms of this model, initially the actual stock of forest resources was in excess of the desired stock. Since the actual stock of existing timber, the old growth, exceeded the desired stock, the economically rational policy was simply to draw down the old growth (actual stock) without any serious consideration being given to the establishment of new stands.[1] In such a world, the initial price of stumpage would be at a very low level, approaching zero, and there would be no economic incentive to invest in tree growing. In fact, the stumpage price was often negative, since the timber resources had no economic value but were merely an obstacle to the use of the land in an alternative pursuit. Gradually, however, an adjustment was occurring. Increases in demand together with a reduction in old-growth stocks brought desired and actual forest stocks into closer relationship. As this occurred, stumpage prices would be expected to exhibit real price increases (Lyon, 1981). The historical occurrence of this phenomenon is well documented. A Resources for the Future (RFF) study completed in the late 1950s (Potter and Christy, 1962) showed a long-run upward trend in the real prices of certain timber resources

---

[1] Although it is often claimed that the old-growth forests were logged without regard to future resource availability and values, a recent study (Johnson and Libecap, 1980) argues that the rate of exploitation of the Lake States' forests was consistent with accurate estimates of future supply and demand.

going back over 100 years. An update of this study in the 1970s (Manthy, 1978) showed no fundamental change in the long-run upward price trend. Although I know of no data that would corroborate this assertion, I suspect that the long-run upward trend was probably worldwide and not confined to North America.

Simultaneously with the occurrence of higher real stumpage prices, technology in tree growing is exhibiting the types of improvements that earlier had been experienced in agriculture. One might hypothesize that such improvements are induced by the more favorable economic situation associated with higher stumpage prices. Whatever the cause of the technological improvements, tree growing technology must still be characterized as "primitive" when compared with that of agriculture. Nevertheless, numerous opportunities for increased yields have been identified, and Hyde (1980) has shown that the economically desired level of technology and management for tree growing in the Pacific Northwest is dependent upon expected future price levels, and that the incremental volumes of timber produced resulting from increased management could be substantial.

The long-term increase in the real price of timber resources and expectations that such price increases will continue, together with advances in the state of technology, appear to be the underlying explanations for the sharp increase in forest investments, including investments in industrial forest plantations that occurred after 1960.

Industrial Forest Plantations

Prior to World War II, there was little interest in investments in industrial forest plantations in most of the world. An exception to this generalization may have been found in Europe and parts of Asia, where forest investments were undertaken. However, for most of the world, the economic costs and expected returns to forest investments apparently did not justify the establishment of commercial forest plantations. The plantations that were established in the United States prior to the 1960s were generally not commercial ventures. For example, man-made forests were established in the 1930s as part of the depression "make-work" programs and in the 1950s as part of the agricultural price support program, for instance, the soil bank.

The experience elsewhere was similar. Although plantation activities had been undertaken in many regions around the globe prior to 1960, these activities were primarily on a pilot basis. By the mid-1960s, however, this situation had changed. The development of improved technologies for planting, growing, and harvesting forests, including genetic techniques to create superior trees, together with the economic incentives provided by higher prices, created a climate that was conducive to investments in forest plantations. By 1975 more that 3 percent of the world's forested area, or 90 million hectares, was man-made forest, although much of this would not be characterized as fast-growing industrial forest (table 2-7). The data further reveal that by 1975 more than 6.5 million hectares of industrial plantations were located in the tropics and the Southern Hemisphere, and Food and Agriculture Organization projections forecast this figure to increase to more than 21 million hectares by the year 2000 (table 2-8).

Major industrial forest plantation activities are under way in New Zealand, Australia, and South America. This trend is particularly strong in South America. Many of the countries on that continent have some form of substantial industrial plantation development under way. In Brazil, some 3.5 million hectares have gone into industrial plantation since the early 1960s, and an additional 300,000 to 500,000 hectares are planted annually (Sedjo, 1980b). Chile currently has more than 700,000 hectares of industrial plantation and is adding to these at about 70,000 per year (Clawson and Sedjo, 1980). Venezuela is developing a vast forest plantation complex on previously unused interior lands. Other activities are under way in other countries.

The implications of plantation forestry are potentially profound. First, forest resource production is becoming and will continue to become more like agricultural crop production. Trees will be planted, grown, and harvested for their wood values. Intensive management, together with improving technology, will increase the productivity of the managed forests. More wood will be grown on less land and by utilizing more inputs. In other words, production will become more intensive.

Second, and perhaps equally as important, the location of industrial forests will gradually change. When old growth is relied on for industrial wood, the critical concerns include the size of the inventory. The inventory was provided by nature in times past, and hence the period of growth required to create the inventory is not relevant to current harvesting decisions. In the future, more and more industrial wood will be provided by

Table 2-7. Area of Man-made Forests, Mid-1970s, by Economic Class and Region

Economic class and region	Million hectares
Developed	
North America	11
Western Europe	13
Oceania	1
Other	10
Total	35
Developing	
Africa	2
Latin America	3
Asia	3
Total	8
Centrally planned	
Europe and the Soviet Union	17
Asia	30
Total	47
Total world	90

Source: Food and Agriculture Organization of the United Nations. Development and Investment in the Forestry Sector. FO:COFO-78/2. 21 (Rome, 1978).

Table 2-8. Industrial Plantations--Tropical and Subtropical America, Africa, and Asia

(thousand hectares)

Region	1975	Projected 1980	Projected 2000
Central and South America	2,786	4,128	10,705
Africa south of Sahara[a]	997	1,248	2,180
Developing Asia[b] and the Far East	2,892	3,719	8,265
Total	6,675	9,095	21,140

Source: J. P. Lanly and J. Clement, "Present and Future Natural Forest and Plantation Areas in the Tropics," Unasylva vol. 31, no. 123 (1979) pp. 17-18.

[a] Excluding South Africa.

[b] From Pakistan east excluding the Peoples Republic of China, Mongolia, and Japan.

plantations whose locations are the result of conscious investment decisions. Time is a critical component of any investment decision, and hence sites and species that permit rapid biological growth will have an inherent economic advantage.

Forest versus Agriculture: Conflicts over Land Use

Traditionally, forests have been the residual claimant on land use. In a world of excess inventories of forests, forests were cut both to obtain the wood resource and to make the land available for other uses. Given that the excessive inventories of naturally created timber utilized vast areas of potential croplands, and given low stumpage prices, forest land use would be expected to give way gradually to other uses. This phenomenon continues today in the tropics, where much concern has been justifiably expressed over the loss of the indigenous forests to pasture and slash-and-burn agriculture. This perspective also continues to exist today in temperate regions. A recent publication estimates that 31 million acres of forest lands have "medium" to "high" potential for conversion to croplands in the United States (USDA and CEQ, 1981).

However, as the excess timber inventories have been absorbed and the relative price of stumpage rises, the extent of the shift out of forest might be expected to abate. This has occurred in much of North America and in Europe. In the South, in New England, and elsewhere in the United States, discarded agricultural lands have found their way back into forests, often as the result of natural regeneration, but increasingly as the result of a conscious afforestation effort. Even in many regions outside of the Northern Hemisphere, forests are making a comeback. Many of the industrial forest plantations of Brazil are occupying lands that had been left infertile due to poor agricultural practices. In Chile, lands are becoming marginal and submarginal for cropping and grazing due to their alternative use in forestry. In New Zealand, plantation forests are beginning to replace sheep-grazing lands, and in Britain there is even an outcry from environmental groups protesting plantation forests replacing moors that were deforested over a thousand years ago.

Tomorrow's Forests

Today the world's woodbasket is filled by a variety of supply sources. As noted, these include natural old-growth forests, naturally regrown forests, artificially regenerated forests created for protection or recreational purposes, and finally, industrial plantation forests. Furthermore, the contribution of the various types of forests to total timber supply is systematically changing over time. The role of old-growth forests is declining while that of industrial plantations is increasing. In addition, the land area in commercially usable forests is changing. While some lands are being taken out of forests and used for agricultural and other nonforest uses, simultaneously other lands are being converted back into forest use.

The extent to which these various types of forests contribute to the supply of industrial wood is largely related to economic considerations. The location of the resource vis-a-vis world markets, its accessibility, the costs of harvest, wood merchantability, and so forth, largely determine the extent to which a forest of any type will be exploited. Therefore, one might expect that certain old-growth forests will eventually be entirely harvested, while other old-growth forests may be only minimally disturbed. In some regions modest artificial regeneration may make economic sense, while other regions may rely upon the less rapidly growing, lower-quality, but less costly natural forest regeneration. In still other regions, alternative uses of the land may preclude further forest use. Finally, on other lands, development of industrial forest plantations may present a viable investment opportunity.

Therefore, at any point in time, the various forests of a large country or of the world as a whole will be in different states of exploitation and development, and they will be making very different contributions to aggregate timber supply.

Eventually, one would expect the transition to be complete and a type of steady-state forest situation to ensue. The long-run, steady-state forest would consist of a mix of industrial forest plantations, artificially regenerated forests, naturally regenerated forests, and, finally, a residual of old-growth forests that are largely unutilized for timber due to their inaccessibility, distance from markets, and lack of merchantability or perhaps of recreational and wilderness values.

THE WORLD'S REQUIREMENTS FOR INDUSTRIAL TIMBERLANDS: SOME SPECULATIONS

The Food and Agriculture Organization (FAO) of the United Nations estimates that the world total forest lands comprise some 2.8 billion hectares (table 2-1). In a separate study, the FAO also estimates that the world demand for industrial roundwood will be about 2 billion cubic meters in the year 2000 (FAO, n.d.). To meet this demand requires an average annual production of less than 1 cubic meter per hectare from each of the world's currently forested hectares. By contrast, a high-yielding industrial plantation produces in excess of 15 cubic meters of roundwood per hectare per year. For pine, yields of more than 20 cubic meters per year are common, and for fast-growing hardwood types, such as eucalyptus, rates of over 25 cubic meters per hectare per year are not unusual. If an overall average yield of 20 cubic meters per hectare per year could be obtained, the entire projected industrial roundwood requirements of the world for the year 2000 could, in theory, be met by just 100 million hectares of high-yielding industrial forest plantation. That is less than 4 percent of the existing forest land of the world. Should the yield of 20 cubic meters per year be deemed impractical for the full requirement of hectares, a yield of only 10 cubic meters per hectare per year on 200 million hectares would meet the projected requirements of the year 2000. That area constitutes only about 7 percent of current world forest lands. Of course, if the technology of tree growing develops as did agricultural technology, the land areas required for industrial forestry would be proportionally diminished.

While the exact amount of suitable land available for high-yielding forest plantations is a matter of speculation, a study undertaken at RFF examined the economics of industrial plantation forestry (Sedjo, 1980a). Nine widely scattered regions in the tropics and Southern Hemisphere with the biological potential for high-yielding plantations were examined, and numerous others were identified. Thus it appears clear that vast amounts of land area are biologically suitable for high-yielding plantations. In addition, the underlying economic conditions appear favorable in all nine regions _provided_ that development costs are not massive and that risk--political, biological, and other--is deemed acceptable. Thus, the limited evidence suggests that considerable biological and economic potential exists, in the long run, for industrial forest plantations.

The foregoing speculations are, of course, not intended as forecasts. Certainly in the near term (at least the next twenty years), existing inventories will continue to provide the bulk of the world's timber supply. Nevertheless, the longer-run implications of the hypotheses above suggest that the potential exists to produce more wood from much less land as plantation forests account for a greater portion of our wood supply.

Long-Term Potential of the United States as a Major Forest Supplier: Some Projections

Historically, the timber production of the United States came predominantly from its vast inventories of old growth. As the country expanded, the forest inventories of New England, the South, the Lake States, the Rockies, and finally the West Coast, were utilized.

Clearly, today the United States has exhausted much if not most of its merchantable and economically accessible old-growth timber as logging activities have marched across the continent. Fortunately, forests are renewable and generally renew naturally. The current second-growth forests of New England, the Lake States, and much of the South are a testimony to natural regeneration. Of course, this naturally regenerated second growth has become part of the nation's forest resource base. In addition, as suggested by the transition hypotheses, the role of artificially regenerated forests and managed forest plantations is becoming increasingly important, for the future the viability of the United States as a source of timber resources will increasingly depend upon the supplies forthcoming from naturally and artificially regenerated second-growth forests and from intensively managed forest plantations.

In recent years two important studies have considered the long-run supply and demand situation of the United States. In the first, an FAO study prepared by a joint forestry and industry working party, a set of worldwide supply and demand projections with the associated trade flows were developed to the year 2000 (FAO, n.d.). These projections indicate an approximate doubling of the U.S. wood deficit between 1960 and 2000, with all of the deficit occurring in the softwoods (see tables 2-9 and 2-10).

In a separate, independent study, the U.S. Forest Service developed a series of fifty-year (1980-2030) projections of trade in timber products as part of its assessment of the forest and rangeland situation in the

Table 2-9. U.S. Outlook to Year 2000 for Wood Products: Softwood (million m^3)

	Consumption	Production	Exports (imports)
1970	214.9	195.6	(19.3)
1980	245.9	220.4	(25.5)
1990	280.7	245.4	(35.3)
2000	315.1	276.8	(38.3)

Source: FAO, "FAO World Outlook: Phase IV," prepared by Joint Forestry and Industry Working Party for the Forestry Department of the Food and Agriculture Organization of the United Nations (n.d.), p. 37.

Table 2-10. U.S. Outlook to Year 2000 for Wood Products: Hardwood (million m^3)

	Consumption	Production	Exports (imports)
1970	63.1	61.8	(1.3)
1980	79.6	78.9	(0.7)
1990	102.9	102.6	(0.3)
2000	135.5	136.4	0.9)

Source: FAO, "FAO World Outlook: Phase IV," prepared by Joint Forestry and Industry Working Party for the Forestry Department of the Food and Agriculture Organization of the United Nations (n.d.), p. 37.

United States (USDA, Forest Service, 1980). This exercise involved disaggregating timber production into various products and projecting trade in each of those products separately. The projections were largely extensions of existing trends with increasing trade deficits by 2030 projected in softwood lumber and wood pulp, while fairly stable trade surpluses prevailed for softwood logs, softwood plywood, and paper and paperboard. Although no aggregate timber volume projections of the type generated by FAO were made, it is clear that the Forest Service projections imply increased aggregate dependence on foreign timber sources (Darr, 1981).

Both of these studies project that large additions to the world's total timber supplies will be forthcoming from Canada, and both imply that much of the increase in the U.S. wood deficit will be met by Canadian timber. However, the ability of Canadian forests to provide these increased timber flows has recently been challenged in a study by F.L.C. Reed (1978).

Two further challenges of a fundamental nature have been directed at the projections of the U.S. Forest Service. First, Zivnuska (1981) notes that the trade projections assume base period prices rather than equilibrium prices. The effect is to inflate domestic consumption, thereby generating a projection bias toward imports. Hence, the validity of the fifty-year trade projections with their large trade deficit is called into question.

Another type of fundamental challenge can be directed at the timber supply projection of the Forest Service. The projections assume that existing forest management practices are essentially unaltered over the fifty-year period. Hence, the projection assumes that forestry will be precluded from experiencing the productivity increases that would be expected in a world of technological improvements and economic incentives (using real stumpage prices) to introduce these improvements.

While it may be inappropriate to generalize to too great an extent from the experience of agriculture to that of forestry, nevertheless if there has been one dominant force affecting agricultural productivity and output, it has been technology (Heady, 1982). To assume, as the Forest Service projections do, that tree growing will experience none of the fruits of technology in the next fifty years is highly questionable. Therefore, in my judgment, the issue of the adequacy of the long-run U.S. timber supply remains an open question.

Industrial Plantation Potential in the United States: More Speculations

With the depletion of old-growth and a growing role for second-growth forests, the ability of U.S. industrial forest plantations to compete with those of the tropics and Southern Hemisphere in major world markets is a critical element in the determination of the future role of the United States as a major source of forest resources and trader of forest products.

The RFF study (Sedjo, 1980a) referred to earlier examines the comparative economic returns to prototype plantations in twelve regions of the world, including two regions of the United States--the Pacific Northwest and the South--as well as the Nordic forests and nine regions in the Southern Hemisphere and tropics.

To generalize, the results of the study suggest that conditions exist in the United States to allow for the economic establishment of forest plantations in both of the regions examined, the South and Pacific Northwest. However, as noted, the results also suggest that the economics appear favorable to the development of forest plantations in all of the nine regions examined in the Southern Hemisphere and the tropics.

Within the United States, the results for the South were particularly favorable for both high-yield sites and average-yield sites. For the Pacific Northwest the results were not as favorable, but nevertheless were acceptable for sawtimber plantations on both high- and average-yield sites, although the average site was near the economic margin. In addition, sensitivity analysis suggested that should the real price of sawtimber stumpage continue to rise, even at relatively modest rates by historical standards, the economic return to the Pacific Northwest sites becomes much more acceptable.

An implication drawn from the study is that the United States has the potential to be an important source of the timber resource even after the demise of its old-growth forest, not only by virtue of the vast land areas available for natural regeneration, but also due to the favorable economics of forest plantations in the United States.

Forest Service data (USDA, Forest Service, 1980) indicate that commercial timberlands in the United States account for about 190 million hectares (483 million acres). However, of this only about 10 percent is the high-productivity class, while another 21 percent is the next productivity class on which high-yielding forest plantations are likely to be economically viable. Together, these site types constitute about 60 million hectares (about 147 million acres). Of this, almost 40 million hectares (100 million acres) is in the South or Pacific Northwest. If the lands of the South and Pacific Northwest are capable of producing an average of 10 cubic meters per hectare per year and if the other features of these lands are such as to provide the acceptable levels of economic return as the results of RFF's plantation study suggest, then plantations on these Southern and Pacific Northwest lands alone have the potential to provide a sustained output equal to about 20 percent of the FAO projected world industrial roundwood requirements for 2000, and about 28 percent of the world industrial conifer requirements. Again, the purpose of this estimate is not as a forecast, but rather as an illustrative exercise to provide some sense of the long-run biological and economic industrial wood potential of U.S. forest plantations. These volumes, of course, would be supplemented by wood forthcoming from regrowth on lower-quality sites.

REFERENCES

Binkley, Clark S. 1979. "Timber Supply from Private Nonindustrial Forests: An Eco-Analysis of Land Owner Behavior," Ph.D. dissertation, School of Forestry and Environmental Studies, Yale University.

Clawson, Marion, and Roger Sedjo. 1980. "Forest Policy." Paper presented at Seminario Economic de Los Recursos Naturales, Santiago, Chile.

Darr, David. 1981. "U.S. Exports and Imports of Some Major Forest Products: The Next Fifty Years," in Roger A. Sedjo, ed., Issues in U.S. International Forest Products Trade. Research Paper No. 23 (Washington, D.C., Resources for the Future).

FAO. n.d. "FAO World Outlook: Phase V: World Outlook for Timber Supply." Prepared by the Joint Forestry and Industry Working Party of the Food and Agriculture Organization of the United Nations.

Heady, Earl. O. 1982. "The Adequacy of Agricultural Land: A Demand-Supply Perspective," in Pierre R. Crosson, ed., The Cropland Crisis: Myth or Reality? (Baltimore, Md., Johns Hopkins University Press for Resources for the Future).

Hyde, William F. 1980. Timber Supply, Land Allocation, and Economic Efficiency (Baltimore: Johns Hopkins University Press for Resources for the Future).

Johnson, Ronald N., and Gary N. Libecap. 1980. "Efficient Markets and Great Lakes Timber: A Conservation Issue Reexamined," Explorations in Economic History vol. 17.

Lyon, Kenneth S. 1981. "Mining of the Forest and the Time Path of Timber," Journal of Environmental Economics and Management (September).

Manthy, Robert S. 1978. Natural Resource Commodities--A Century of Statistics (Baltimore: Johns Hopkins University Press for Resources for the Future).

Potter, Neal, and Francis T. Christy, Jr. 1962. Trends in Natural Resource Commodities (Baltimore: Johns Hopkins University Press for Resources for the Future.

Reed, F.L.C. and Associates, Ltd. 1978. Forest Management in Canada. Vols. I and II. Forest Management Institute Information Report FMR-X-102, Canadian Forestry Service, Environment Canada.

Sedjo, Roger A. 1980a. "U.S. Comparative Advantage in Timber Growing." Paper presented at Forest Products Research Society Meeting, "Timber Demand: The Future is Now."

_____. 1980b. "Forest Plantations in Brazil and Their Possible Effects on World Pulp Market," Journal of Forestry (December) pp. 702-705.

_____, and Samuel J. Radcliffe. 1981. Postwar Trends in U.S. Forest Products Trade: A Global, National, and Regional View. Research Paper No. 22 (Washington, D.C., Resources for the Future).

U.S. Department of Agriculture and Council on Environmental Quality. 1981. National Agricultural Lands Study (Washington, D.C.).

U.S. Department of Agriculture, Forest Service. 1980. An Analysis of the Timber Situation of the United States: 1952-2030. Review draft (Washington, D.C.).

Zivnuska, John. 1981. "Discussion," in Roger A. Sedjo, ed., Issues in U.S. International Forest Products Trade. Research Paper No. 23 (Washington, D.C., Resources for the Future).

Hans Gregersen

COMMENTS

The following comments focus on three interrelated conclusions put forth in chapter 2, which outlines a number of factors that could affect pressures on U.S. forest lands in the future.

First, Roger Sedjo concludes that there is an accelerated global shift away from reliance on natural forests and toward dependence on high-yield plantation forests, many of which utilize sophisticated technology. Further, he suggests that industrial forest plantations will tend to gravitate toward lands that have high productivity and are economically well situated and that these lands may often be in regions different from those that have traditionally had large natural old-growth stands.

Second, based on available statistics, Sedjo concludes that the nontraditional producer countries (mainly in the tropics and the Southern Hemisphere) are increasing their share of total world production of both industrial wood and other roundwood.

These two shifts have direct implications in terms of a third conclusion. Sedjo states it as follows: "The ability of U.S. industrial forest plantations to compete with those of the tropics and Southern Hemisphere in major world markets is a critical element in the determination of the future role of the United States as a major source of forest resources and trader of forest products."

The first two conclusions are based on facts as we know them. The third one I basically agree with, if we look twenty to fifty years into the future. However, in the shorter term there are a number of factors that will, with regard to the United States, act to dampen the impacts of the global shifts Sedjo discusses.

I will briefly mention some of these factors and then suggest one additional internationally related influence on U.S. forest lands not discussed by Sedjo.

DEMAND FACTORS AND USE OF PLANTATION WOOD

A first factor that will dampen impacts on the United States from global shifts relates to markets and uses for wood. In discussing the impact of industrial plantations in other countries on U.S. forestry, we need to consider very carefully the demand side and intended use for the expanding plantation outputs. In some cases such uses do not relate at all to the U.S. forestry situation.

One example will illustrate this point. In terms of rapid growth of plantation area, Sedjo refers to Brazil with a current industrial plantation area of some 3.5 million hectares and an additional annual planting rate of some 300,000 to 500,000 hectares. There is no denying that these are impressive figures and imply substantial wood production increases. However, we need to temper these figures with an understanding of the intended use for their output.

A recent study of Brazilian wood energy points out that about one-third of the plantations in Brazil are for charcoal production and will not be providing forest products that compete with the United States. Further, looking at the Brazilian government's goal for charcoal-blown pig iron, we find that some 300,000 hectares per year of new plantations would be needed by 1985 to meet the target (Beattie, 1979). The basic point is that all the new plantation wood will not be available for traditional wood products that could compete with U.S. production.

Worldwide, the question of fuelwood is of great importance. I do not think that many of us have fully comprehended the scale of the present world fuelwood crisis and the intensity of likely future political pressures to divert forestry efforts to fuelwood. A comparison will provide some perspective on the relative magnitudes involved. Sedjo mentions the Food and Agriculture Organization projection which estimates that by the year 2000 there will be 21 million hectares of industrial plantations. On the other hand, estimates from the World Bank indicate that some 50 million hectares of _additional_ fuelwood plantations would be needed by the year 2000 just to meet additional expected fuelwood requirements. While it is

difficult to envision that this level of planting will take place, it is
not difficult to envision that the pressures in various countries to move
more rapidly in the fuelwood area could divert forestry efforts and lands
from industrial plantations.

Quite aside from the fuelwood question, any discussion of the effects
of global plantation and production developments on the United States has
to be tempered by a look at demand for traditional wood products in the
producing countries. While plantation establishment and wood products production are increasing more rapidly in the nontraditional producer countries, so is the demand for forest products in some cases. The result is
a widening gap between production and consumption and may mean an expanding
market for exports from traditional traders such as North America.

Ward (1981), in a Resources for the Future conference on Issues in
U.S. International Forest Products Trade, indicated that for many products,
such as softwood lumber, softwood plywood, and hardwood lumber, the expected growth in world demand and world import demand will be considerably
higher than growth of U.S. demand as projected by the Forest Service. Figures for individual countries experiencing the most rapid growth in either
plantation establishment or wood product output increases or both provide
similar evidence.

Korea is one of the shining examples of a country that has expanded
its plantation area very rapidly. Between 1973 and 1978 alone, Korea established over 1 million hectares of plantations. Another 1.5 million
hectares are planned for the 1978-87 period (Office of Forestry, Republic
of Korea, 1981). Yet despite these major efforts and Korea's desire to
become self-sufficient in wood, official estimates indicate that Korea
will only have increased from 11 percent self-sufficiency now to about 28
percent by the turn of the century. Even by the year 2030, the planned
level only rises to 50 percent (Office of Forestry, Republic of Korea,
n.d.). Obviously, Korea will be importing wood for many years to come.

As somewhat of an aside, Korea is also an interesting example of a
wood-scarce country that is a net <u>exporter</u> of wood products in value terms
(Lee, 1981). The United States is just the opposite--a wood-rich country
that is also a net importer of wood products. As Sedjo points out, there
are a number of factors other than wood availability that determine a
country's net trade position in forest products.

AVAILABILITY OF INVESTMENT CAPITAL

A second factor dampening the potential rate of expansion of output and exports from the nontraditional producers is availability of investment capital and technical/managerial expertise. Given current attitudes, a shift in log export limitation policies is not likely to occur. Thus, investment in domestic processing is the order of the day if exports are to expand.

Foreign investment will contribute to the expansion and has done so in the past. U.S. companies have traditionally been heavily involved overseas. For example, about 50 percent of Latin American pulp and paper sales in 1966 were accounted for by U.S.-owned companies and their affiliates. Further, almost fifty U.S. companies, including all the major forest products firms, had investments in Latin America in 1975 (Gregersen and Contreras, 1975).

Even with an optimistic view of foreign investment, the capital requirements to meet just future domestic needs in the nontraditional producing countries appear to be substantial. It has been estimated conservatively that some $3.2 billion 1979 dollars would have to be invested <u>annually</u> for Latin America to achieve regional self-sufficiency in forest products by 1994 (Contreras, 1980). Export expansion and competition with the United States in its traditional markets would require additional capital.

At this point we should hasten to emphasize that the above comments about dampening effects should not be taken to imply that there necessarily will be a never-ending future market for North American wood products; nor are there necessarily unlimited opportunities for expansion of U.S. investment in the forest-based sector in the tropics.

There will likely be a steady if not dramatic increase in the world market for wood products. To some extent, what Sedjo calls the nontraditional producers will move into the market, and they will certainly become more self-sufficient as their industrial capacity increases. While some increases in U.S. exports are possible, there is no reason to foresee the United States becoming the "woodbasket" of the world in the way that people talk about the United States becoming the "breadbasket" of the world.

As a matter of fact, there are some significant potential conflicts between the breadbasket and the woodbasket. Sedjo talks about forest

forest versus agricultural conflicts over land use. In chapter 2, Clawson points out that over the past couple of decades there has been a fairly stable relationship between amount of commercial forest and of farmland, that is, with no major shifts either way. However, if the United States is to increasingly become the foodbasket of the world, there will be mounting pressures to find new cropland. In anticipation of such future needs, a recent U.S. Department of Agriculture publication suggests that about 31 million acres of forest lands have "medium" to "high" agricultural crop potentials (USDA, Soil Conservation Service, 1980). Obviously, expansion of U.S. agricultural exports could put some significant pressures on U.S. forest lands.

INTERNATIONAL TECHNOLOGY TRANSFER

In addition to the points discussed in chapter 2, there is one other significant interaction between the United States and other countries that has had profound impacts on how we cope with pressures on U.S. forest lands. International technology transfer is often ignored or considered only in terms of transfers from developed to developing countries. In point of fact, technology transfer between developed countries has had profound influences in terms of species and log-size utilization and forest products trade.

For example, U.S. adaptation of particle board technology from Europe permitted us to utilize wood resources that had previously not been used. As previous "weed" species became useful, pressures on U.S. forest lands shifted. Similarly, technology imported from Finland permitted U.S. producers of plywood to effectively utilize smaller-diameter logs. These and many other imported technologies have in effect resulted in reduced pressures on certain U.S. forest types and increased pressures on others (for example, the remarkable change in attitude toward aspen lands in Minnesota due to importation of improved wafer board technology from Canada).

Similarly, the United States has also exported many significant technologies, both in forestry and forest products. These technologies have benefited other countries, in some cases resulting in reductions of their imports from the United States or increasing their exports and thus again reducing pressures on U.S. forest lands.

Future technology transfers could be equally as important. For example, if the U.S. Forest Products Laboratory's new press-drying technology for paper becomes properly diffused and developed, it could have profound impacts in countries short on energy and long on hardwood resources. This in turn could affect U.S. trade and, eventually, pressures on U.S. forest lands.

Based on research carried out at RFF, Sedjo concludes that the evidence suggests that forest plantations in the United States can probably compete with the fast-growing plantations of the tropics and the Southern Hemisphere. The question remains: Will the United States gear up to compete by developing appropriate policies to encourage forest products trade?

REFERENCES

Beattie, W. 1979. "Energy Production from the Brazilian Forestry Subsector." Draft report prepared for the World Bank, Washington, D.C.

Contreras, A. 1980. "Evolution of Requirements for Forest Products in Latin America," Forest Products Journal vol. 30, no. 10, pp. 70-74.

Gregersen, H. M., and A. Contreras. 1975. U.S. Investment in the Forest-based Sector in Latin America (Baltimore, Johns Hopkins University Press for Resources for the Future).

Lee, Phil Woo. 1981. "Timber Demand and Forest Products Trade," Forest Products Journal vol. 31, no. 1, pp. 12-13.

Office of Forestry, Republic of Korea. 1981. National Progress Report on Korea, 1977-1980 (Seoul, Office of Forestry).

_____. n.d. "Synopsis of the Second 10-Year Forestry Development Plan, 1979-1988" (Seoul, Office of Forestry).

U.S. Department of Agriculture, Soil Conservation Service. 1980. America's Soil and Water: Condition and Trends (Washington, D.C.).

Ward, J. 1981. "Discussion" in Roger A. Sedjo, ed., Issues in U.S. International Forest Products Trade. Research Paper No. 23 (Washington, D.C., Resources for the Future).

PART II

PUBLIC INTERVENTION ON PRIVATE FOREST LANDS

OVERVIEW OF PART II

A variety of questions can be addressed under the heading of public intervention on private forest lands: Under what conditions is such intervention socially justifiable in the context of a market economy in which private ownership of property is a critical component? What forms does regulation take when applied to forest lands? Is regulation as practiced actually in the broad public interest? Does intervention result in a mix of forest outputs that is more optimal than would be produced in the absence of regulation? What are the trade-offs between regulation and the direct government "taking" of land? Given our constitutional protection against taking without compensation, the government finds it advantageous to regulate some land into uses it deems appropriate without actually acquiring ownership. Is such action socially desirable? What are the implications of such regulations as regards the asset value of property? If asset values do decline, does this constitute taking as protected by the Fifth Amendment?

These major questions are considered in the context of the issues addressed in part II. Specifically, in chapter 3, Sterling Brubaker discusses land use concepts and articulates the rationale for government intervention in private land use. In chapter 4, Henry Vaux describes the experience of California in regulating its forest lands. And finally, the taking issue and its economic and social implications are addressed by Bruce Johnson in chapter 5.

Public intervention in private land use in a market economy is usually justified on an economic basis in terms of some type of market failure.

In the absence of a market failure, the private owner will have a set of economic incentives to utilize the land to generate the social optimal level and mix of outputs through time. The private owner will also have economic incentive to practice appropriate measures to maintain the fertility and productivity of the land, since a perceptible deterioration in productivity will be reflected in the current market value of the land.

However, it is universally acknowledged that market failures do exist. Most market failures can be viewed as some type of common property problem. In the absence of clear property rights, one would generally expect the rate of utilization of a resource to exceed the social optimum. In the classical example of the "tragedy of the commons," a commonly owned pasture is overgrazed because no one has the right to limit grazing to the optimal number of sheep. Certain individuals experience the benefits in the form of the growth of their privately owned sheep, while the costs--in terms of pasture deterioration--are borne jointly by the "common" pasture ownership. Thus an externality is generated, that is, the costs are borne by a group that is external to those receiving the benefits. Similarly, a commonly owned forest may be overhunted, the commonly owned waterways passing through and beyond the forest may suffer from excessive silting, thereby generating downstream costs, and the commonly enjoyed visual amenities may come into conflict with the privately generated benefits of a timber harvest.

THE RATIONALE FOR INTERVENTION

In chapter 3, Brubaker develops the concept of land as a resource and as property. He examines the social objectives related to the land, the conflicts between private and social returns, and a rationale for public intervention. The rationale for intervention "is that it enhances real wealth or social welfare in ways that the market is unable to accomplish," or, in other words, it provides the basis for a more socially optimal mix of outputs. In addition to discussing the usual external effects, the author presents the view that unregulated private land will not adequately ensure the welfare of future generations. Brubaker maintains that private owners often have insufficient incentive to prevent environmental damages adequately or to preserve unique and unusual ecological systems and genetic information.

In his comments on chapter 3, Richard Stroup raises three fundamental types of objection to Brubaker's view. First, Stroup argues that the incentives generated by the market system will, in fact, adequately address many of the concerns raised by Brubaker. For example, he maintains that the market generally provides proper economic incentives to maintain the socially desirable level of protection of the fertility and productivity of the land resource, since the costs of the loss of productivity are reflected in the market price of the land and hence are borne by the private owner. Second, Stroup exhibits a much greater degree of skepticism about the ability of the political or judicial processes to generate intervention that will result in socially desirable outcomes, even where markets are operating imperfectly. The political process, he argues, is more myopic than is the market. Finally, in Stroup's view, much of the difficulty associated with the lack of incentives for the preservation of unique and unusual systems is due to the lack of property rights for some of these features. Therefore, the problem of preserving unique resources is more often due to inappropriate institutional agreements than to the inability of markets to function properly.

STATE INTERVENTION

The history and implications of public intervention in the use and management of private property, and specifically the forest lands in California, are presented by Vaux in chapter 4. This case study deals with a state whose system of regulating forest practices is one of the most comprehensive and stringent in the nation. After describing this system Vaux analyzes its influence on current and future levels of output and its effect on incentives for investment. Finally, he evaluates the cumulative effect of the various interventions and the achievement of the state's statutory goals in forest policy.

The numerous interventions in California include protections, taxation, financial aid to encourage research in the rehabilitation of forest land, and professional regulation of timber harvesting practices. An important regulation includes the provision that timber will be cut only in accordance with an approved timber harvest plan.

In his evaluation of the impact of the intervention, Vaux concludes that forest protection has been increased; that financial aid programs have helped encourage reforestation, particularly among small owners; that education, research, and analysis have been furthered; and that the yield tax "reduces the owner's actual cumulated tax burden at the end of the cutting cycle."

In assessing state regulation, Vaux concludes that one effect is likely to be greater future timber supplies and a resulting decrease in timber prices to California consumers. With respect to the achievement of the statutory goals for environmental protection, the author notes the difficulty of evaluation because of the impreciseness of the goals, but he deduces that the good record of compliance suggests increased environmental protection. Vaux estimates that additional cash expenditures by California's forest owners due to compliance with the logging regulations are about $55 million per year, and he suggests that these costs vary by ownership. Finally, he addresses the broad issue of the effects of regulation such as that in California on private investments in timber growing. However, Vaux points out that a definitive determination of these effects is most difficult, and he surmises that regulation may reduce the level of investment by the forest industry while increasing it for nonindustrial private owners.

In his critique of chapter 4, William Moshofsky takes basic issue with the concept that regulators should be allowed to impose costly restrictions on land use without the regulatory agency's being required to reimburse landowners for the cost of compliance. The difference between forest land use and other land use regulation, according to Moshofsky, is that the purpose and effect of the former is not primarily to prevent harm. Rather, it is designed to redistribute benefits from one group to another. (This issue is examined in chapter 5.)

Moshofsky also argues that many of the regulations in California and elsewhere are not related to bona fide economic externalities. Reforestation concerns, he argues, have little to do with types of environmental concerns for which there are recognizable externalites. By contrast, Moshofsky finds justification for intervention in water issues, because the water in streams and rivers _is_ publicly owned and externalities _are_ present. This rationale, however, does not provide the basis for extending the regulation to adjoining lands.

In summary, Moshofsky accepts government intervention as a way to prevent harm in the form of externalities and as a way to manage publicly owned resources such as water. However, he is concerned with intervention that exceeds this rationale and infringes on private property rights, is used primarily to facilitate redistribution of wealth from land owners to others, and reduces or eliminates value without compensation.

THE TAKING ISSUE IN FORESTRY

The Fifth Amendment to the U.S. Constitution contains the famous provision that private property may not be taken for public use without just compensation. This provision apparently protects private property rights and guarantees "just compensation" should property be taken to further the "public good." But what is to prevent government from regulating land into certain uses without a transfer of ownership, and thereby controlling the use of the land without having to compensate owners for economic losses incurred. Must owners bear economic losses in such a situation? Economic reasoning would indicate that often they do. The value of an asset, including real property, is related to its ability to generate economic returns in the form of current and future income streams plus appreciation. To restrict or prohibit certain economic use affects the current and future income streams by reducing the present net value, that is, the market value, of the asset. Thus, a land use restriction that reduces income constitutes, in an economic sense, the taking of value.

A controversy over the constitutionality of such taking has been raging in the courts and legal literature, with its most famous defense made by Bosselman, Callies, and Banta in The Taking Issue.[1]

Obviously, forest owners as landowners are potentially affected by such taking. As discussed in chapter 4, California practices a number of interventions. Timber harvesting practices are regulated, and a harvest plan can be denied if it does not conform to the rules of the forestry board.

---

[1] Fred Bosselman, David Callies, and John Banta, The Taking Issue: An Analysis of the Constitutional Limits of Land Use Control. Council on Environmental Quality (Washington, D.C., Government Printing Office, 1973).

Examples of regulation that might result in taking from private forest owners include the prohibition of clear-cutting, the requirement of protected unharvested timber margins alongside streams, the prohibition of cutting alongside roads to protect scenic values, the legal requirement of regenerating timber stands, and so forth. To the extent that each of these practices would not be undertaken voluntarily, the owner's net private earning stream from the forest resource is reduced, and economic theory would predict that the asset value of the forest land would be negatively impacted.

However, the net social effect resulting from regulations such as those listed above may be different from the private effect. For example, while the requirement of protection stands along a stream may lower the private value of the forest land asset, it can preclude or reduce the creation of costly downstream effects, and therefore may increase social welfare. This type of regulation may also be justified as a method for shifting the downstream costs of timber harvest to the forest owner. It may or may not, however, be an efficient approach for dealing with downstream effects. Even if it is efficient, the lack of compensation puts the costs of stream protection directly on the forest landowner. Should the owner be forced to bear these costs? Alternatively, does the landowner have the right to pollute a stream or to engage in practices that will result in costs to others? A similar but slightly different situation may arise if, for example, a forest margin is required along roads solely for the sake of preserving visual amenities for travelers. In this case the income forgone by not harvesting trees is a measure of the costs associated with preserving visual amenities. In the absence of compensation, the costs of the the preservation would again fall on forest landowners, while the visual amenity benefits would accrue to travelers on roads through these forests. One can argue the equity of such an approach.

With regard to both of the situations just mentioned, two important issues arise. First, who should bear the costs, the forest owners or the water/amenity users? If the unharvested forests are creating external water/amenity benefits, would fairness dictate that the forest landowners be compensated for these services? If no compensation is forthcoming, should the forest landowners be forced to provide these services in perpetuity? What are the rights of private ownership to these harvests?

The second issue is one of economic efficiency. Are the benefits associated with regulating the harvest along streams and roads justified in terms of the benefits forgone by reducing the harvest? Is this the cost-efficient manner of preserving those benefits? Might a lower (or higher) standard of preservation of water quality and amenities be economically justified?

In chapter 5, Bruce Johnson points out some of the difficulties encountered in the morass of economic, political, and legal intricacies related to the externalities and the taking issue. Johnson notes that several avenues of recourse are available to those who believe they are injured from undesirable market spillovers. They could, in theory, (a) use the contract process, (b) purchase property rights, (c) resort to force, or (d) enlist the government to intervene in their behalf. Since political activity via government regulation is the least costly alternative, it is not surprising that this approach is most commonly practiced.

Johnson discusses recent government and judicial action regarding the taking issue. For the most part, he says, compensation is not forthcoming. Johnson argues that the trend toward increased regulation has grown stronger and suggests that it is likely to continue. The author ignores the usual economic efficiency arguments, noting that, as a practical matter these arguments do not appear to be persuasive with politicians and judges. Instead, he attacks the problem from the perspective of political economy with its emphasis upon equity and redistribution. He argues by illustration that public hearings tend to be biased and partial.

Johnson maintains that the existence of spillovers, which by definition imply a degree of market failure, does not imply "the heroic assumption that the existing regulations are operating at the appropriate scale or that the costs and benefits of such programs are appropriately distributed among landowners." He perceives forest land disputes not principally as disagreements concerning the externalities of timber harvest, but rather as a battle over the issue of using forest land to minister "to the public's 'need' for natural vistas and wildlife habitats, rather than timber production." Thus, the implication is that intervention does not improve the management of the resource from the societal viewpoint, but rather reallocates the output according to the preferences of one group instead of another.

In his comments on chapter 5, Bosselman focuses on the legal justification of the Fifth Amendment as it applies to regulation. He argues that land use regulation precedents abounded in colonial and postcolonial periods of the United States even though land use regulation was often extreme. Thus, Bosselman maintains that the historical precedents suggest that the Founding Fathers did not intend that the Fifth Amendment requirements of compensation be applicable to the taking that occurs as the result of regulation.

Chapter 3

LAND USE CONCEPTS

Sterling Brubaker

Land as a resource has highly complex attributes and capabilities, most of which can be organized according to economic principles for present purposes. Land as property carries rights that may either facilitate or complicate the provision of its services. The purpose of this chapter is to discuss some of the characteristics of land as a resource and as property. What does land do? What do we want of it? What is the role of the market in satisfying those wants, and where is public intervention called for? The following discussion focuses on private rather than on public lands, but many land use problems are common to both ownerships. Although the chapter does not deal with forest land in particular, it applies in part to forest as well as to other lands.

LAND AS A RESOURCE

Land is the resource with which human beings have had their most intimate relationship throughout the history of the race. It has provided most of our sustenance and many of our pleasures. While primitive people took nature as they found it, as shepherds and tillers they began a fundamental alteration in the use of land. On the basis of flocks and settlements, they constructed the social system that has fostered our ecological primacy and made us the most profound influence on the natural system. With agriculture, our dependence on the land and attachment to it became even stronger.

Against that background came more recent changes. As industrial, scientific, and economic people--a very recent development--we have increased our power over nature and greatly increased our numbers while becoming more and more estranged from the land--we are now very substantially a race of miners, dependent on depletable resources. Although direct dependence on land and earlier attitudes toward its religious and social meaning have weakened, they have not vanished. Now veiled by a complex legal and economic system, land nonetheless remains important to us, and ancient attitudes resist stuffing land neatly into the economist's baggage.

Land is generally viewed as a renewable resource. Some of its attributes are essentially indestructible--for example, surface areal extension. But in specific uses land may not be renewable, or may be so only if managed with that objective in view. With a given technology and level of use, perpetual use often is possible; with other technology or levels of use, temporary or permanent degradation may limit the capacity of land to yield specific services.

Most uses of land involve its transformation to some degree. As the intensity of use or degree of transformation rises, reversion to an earlier state becomes technically and economically very difficult, if not impossible. Most often, reversals signify disaster rather than progress-- the jungle reclaims Mayan temples, the forest covers the abandoned Appalachian cornpatch, and so forth.

Excluding subsurface resources from consideration, land as a resource can be used to grow things, to provide places for human infrastructure, to yield environmental services and amenities, to nurture ecological and genetic preserves, and to provide collection areas for direct solar energy and water. Of course the categories can overlap, since amenity values can be associated with many other uses. Furthermore, a given parcel of land often may produce joint or multiple prod cts. The same parcel often is versatile enough to serve several alternative primary uses, and in each of these uses it may yield perpetual services of that sort or be able to accommodate a more intensive use. All of these possible uses of land are of value, whether or not they are marketed.

As part of the cycling of global systems, land provides environmental services. At the local level, land buffers natural discharges and assimilates humanly generated wastes. Often these services occur without human intervention, although we can modify the effectiveness and completeness of such functions by action on the land. For instance, acid rain absorbed and filtered by the soil is partially neutralized in a natural way, whereas watershed management can intervene to affect the rate of runoff and the sediment load carried by runoff. Wetlands management can affect the assimilative capacity of those lands or the ecological support system they offer to desired species. In other cases--for example, sludge disposal or landfill sites--there is direct intervention to produce environmental services from the land.

Environmental amenities are also provided by both natural and modified systems, although their character is different in each of the two cases. Open space, for example, becomes an amenity whose value is created by surrounding congestion. The amenity value of wilderness, while enhanced by crowding elsewhere in the society, is degraded if that crowding intrudes too closely.

Land also provides ecological and genetic preserves, which may have amenity or other commercial value but are principally a form of insurance against the irretrievable loss of life forms of potential future value to human beings. Beyond that, they may escape conventional human measures of values. Our bias against the willful extermination of species does not appear to be based on any economic calculus.

The principal use of land in the United States is for the production of agricultural and forestry commodities, which together employ more than 70 percent of the surface area (USDA, 1973). To be suitable for these purposes, land must have characteristics of soil, climate, and relief that permit profitable production with available technology and current price expectations (USDA, 1975). If land is suitable for producing more than one branch of commodity, it tends to move into the most profitable branch that market conditions allow.

While total land area is fixed, its allocation among use categories is not, and so there is competition for land among uses. As discussed in chapter 1, crop use can ordinarily attract rural land suitable for crops away from pasture, range, or forest when demand for crops is strong. Land frequently moves between crops and pasture in response to demand fluctuations. Growth in the long-term demand for cropland favors the conversion of forest and pasture to crops so long as the new use promises to yield more than the opportunity costs of the old plus conversion costs. At the same time, some cropland is continually lost to more intensive uses.

Forest lands stand near the bottom of the hierarchy and can seldom displace other uses, although regional shifts in crop production have idled some former cropland in the East and South, with much of it reverting to forest.

As previously noted, commodity production can be conducted so as not to damage the productive capacity of the resource. It also can be conducted so as to produce noncommodity benefits, such as watershed protection, wildlife, or even agreeable scenery. However, it may not be profitable to do either of these. Thus, a potentially renewable resource can be degraded, and the production of positive externalities may be neglected. The erosion of cropland is probably the most pervasive case of conflict between current returns and the long-term preservation of the resource or control of externalities, in this case, negative ones.

While commodity production employs the largest areas, land use for purposes of human infrastructure confers the most economic value on land and leads to some of the most acute conflicts over use. There are advantages in conducting many urban pursuits close to others. Downtown is where economic actors cluster and thus sites at the metropolitan core command a premium. Land values diminish away from the core, although they rise again around outlying transportation nodes or adjacent to desired amenities. Still farther out, they tend toward the value of the land in commodity production. Land in virtually any urban or infrastructure use (including housing) is more valuable than the same land used for agriculture or forestry. As a consequence, urban use makes first claim on land whenever the market governs. Restrictions on the supply of land available for urban use or for its more intensive variants raise the price of urban land still further.

Conflicts arise over which land will be privileged through zoning to enter urban use, or its more intensive categories, and over public investments that may enhance some values but not others. Conflicts also arise over incompatible uses in close proximity and over public investments that threaten to diminish existing property values. Thus, the new highway that passes through my neighborhood is a bane if too close, but just fine if a nearby interchange makes my land more valuable. Being at the peak of the hierarchy of uses, land taken into urban use tends to remain there. The location is valuable and the investment in improvements and utilities favors continued urban use.

The foregoing discussion of land as a resource is without regard to ownership. It is in the interest of society to have land serve its various functions well, no matter who owns it. However, ownership does affect what kind of services land will provide and how it will be maintained.

LAND AS PROPERTY

Legally, land in the United States is either private or public property, but not common property. Public land is public property and is subject to whatever rules the public law provides, but private citizens do not have unrestricted access to it.

Private lands enjoy the constitutional protection that other forms of property also receive. Owners of property are entitled to its exclusive use, can sell or will it to others, and can earn income from its productive use. Owners at one time were accorded the right to use their property very much as they chose, and in the popular view this remains a right of property ownership.

However, the right to use property (and hence its value) is now somewhat more circumscribed. Nuisance doctrine always limited the use of property. More recently, the use of real property has been regulated by a number of devices--environmental protection, social rules (for example, rules against racial discrimination), economic regulations (such as rent controls), and, above all, planning and zoning. All of these can dramatically affect value. While leaving intact the other rights of property, including the theoretical right to use property to the owner's best advantage, regulation has made it difficult to realize many potential gains. Nonetheless, public actions of the sort mentioned can avoid the legal proscription against public taking of private property if some uses are still permitted, despite the impairment of much of the value of the investment.

Public action also can increase the value of property in many ways. For example, lifting of previously imposed restrictions or the passage of more generous zoning rules can give a windfall to current owners. Public action in other fields also may have special effects on property values. Investment in utility or transportation enhances the value of undeveloped land. And broader economic policies that favor ownership (such as tax benefits, credit guarantees, or lower interest rates) also can increase property value. There is strong incentive to secure these changes, and in some instances they may be won by the influence of either persuasion or corruption. While public subsidies and favorable rules are a major source of increased value on specific properties, they are not a unique source. A private utility extension for which the owner pays full cost may more than pay for itself in increased property value. Ultimately, social growth and development are required to validate increases in local urban property values.

The bias of our legal and economic system is to accord strong rights to property and to require good reasons if those rights are in any way limited. The legal bias has a long history in the law and is explicitly grounded in the Constitution. The bias in economics derives from the assumption that resources, including real property, will be drawn into their most profitable uses and will lead to efficient use of resources if owners are free to employ them as they choose. Efficiency in this case is relative to an existing distribution of income according to which preferences for resource use are established. Within that limitation, to the extent that prices are competitive and fully reflective of costs and benefits, owners' decisions about use of their property are apt to lead to efficient allocation of resources and thereby to maximization of the wealth of society. Critics point to various failures of the system to attain this ideal, but belief in it as a tendency is a powerful support for the rights of property. These rights are reinforced by considerations of equity that hold that owners acquiring property in good faith should not be deprived of it by subsequent social action. By extension, the right to use property, which gives it its value, is presumed in the absence of strong counterarguments.

SOCIAL OBJECTIVES

What do we as a society want from the land? Is there any difference between what we want from private and public lands? In principle, it appears that we want the same from both--that is, that they should be used so as to maximize net benefits, subject to the constraints we put on that objective. In the aggregate, public and private lands differ in their physical and economic characteristics, and so the resource services that they yield differ. (For example, public lands have scant acreage adapted to crops or favorably situated for urban use.) For for both public and private lands, we ask that they produce those services that yield the greatest net benefits.

The economic criterion is of broad embrace. It incorporates the public goods[1] and externalities that may escape private reckoning. Thus what society wants from private lands may diverge from what private owners want, since the latter are concerned with private maximization and not with the greatest benefit to society. In particular, private lands will not produce the level of public goods that they are capable of because owners have no private incentive to seek that goal.

As a society, we feel a longer-term obligation to the future than is apt to be the case for private owners. Admittedly our political instruments for expressing this interest are disastrously shortsighted, but we think of society as continuing indefinitely, and we recognize an obligation to leave future generations with a relatively intact resource base. The main resource that we aspire to leave them is improved technology that grants access to lower-quality natural resources on a basis equal to the present. In the case of land, however, we have the option of leaving the physical resource intact, and, in some of its uses, there may be no very good alternative to this option if we are to give future generations equal access. Thus, preservation of the productive capability of the land becomes a social objective even where it may not be an objective for a private owner.

Society has other objectives that often are expressed through land use controls but that are not strictly land use issues as viewed from the resource standpoint. These issues may have great implications for land as property. Policies to discourage racial discrimination, to protect free movement, or to encourage greater sense of community may operate through land use controls or prohibitions on land use. However, the same objectives could in most instances be sought through non-land-use approaches. For example, the goal of assisting needy renters can be accomplished at least as well through rent or income subsidy as through rent

---

[1] "Public goods" is a technical term referring to those goods or services that if provided for one are provided for all and that if used by one are not diminished in the amount available to others. Hence they are best made widely available.

control. Land use controls have provided a convenient device for pursuit of diverse objectives, but they may not be the most efficient means for doing so (Erwin and coauthors, 1977).

RATIONALE FOR PUBLIC INTERVENTION

The public may intervene in land use decisions directly or may influence them by incentive. Such action is controversial in most cases because of its effect on property values. The rationale for intervention usually is that it enhances real wealth or social welfare in ways that the market cannot. Intervention may seek to minimize unfavorable externalities to ease the cost of public services, or to provide public goods, and it may also seek to secure the interests of future generations. As noted, there may be other objectives of a social or economic character pursued through land use policies. Let us examine the hoped-for economic benefits in more detail.

Land often has the capability of producing services in the public goods category. Because such goods, though valuable, have no market, they may not be made available from private land even when they can be produced at little cost. Free passage for birdwatchers is one example. Public goods may still be produced on private land if they are without cost or are a joint product of other marketed goods, but there is no reason to expect output to be at the optimum level.

Public services—sewers, water, electricity, gas, transportation, schools, libraries, and so forth—are more efficiently provided in reasonably compact settlements. If the cost of providing them was reflected in price to users, then users could make rational decisions about location that incorporated such costs. In fact, many public services are unpriced or are priced in ways that do not reflect the cost of provision to new users. This favors dispersion and increased cost of services and also means that more land is impacted by urban use and withdrawn from other productive use. Intervention that diminishes dispersion—for example, marginal-cost utility parking or effective zoning—helps correct this loss.

Externalities are those costs and benefits that are not borne by the one using resources and hence that do not enter into that person's decision to produce or consume. Environmental damages are the commonest example. Such damages can be reduced by land use controls in many cases.

An industry causes less damage if located far enough from population centers to allow dispersion and assimilation to reduce its noxious effects; or concentration of some waste streams, such as sewage, may permit treatment and relief of environmental burden. In commodity production, the way in which the land is operated strongly affects the environmental burden created. Intervention to control externalities may or may not have a land use component, but the control of land use may be an effective ingredient of some types of control.

Securing the interests of future generations (beyond one or two) is a social and not a private concern. Ordinarily it is beyond the ability of a private owner to do much about genetic or ecological preserves, for example. Nor can a private owner ensure that adequate farmland or forest will remain for the future. Even worse, operating the owner's land may not prove profitable enough for that person to maintain its productive capacity. If there is a social interest in these matters, it can only be protected by public intervention; private individuals lack the power or incentive to do it.

Apart from its effectiveness in achieving its primary purpose, the main issue presented by public intervention is always the question of equity. There are gainers and losers from almost any social action, but actions affecting land values are compounded by the attitudes toward and the constitutional protection accorded real property. If intervention accomplishes its objectives and increases net welfare, to whom does the increase belong? Does ownership of property give one a prior claim? Should owners bear costs not asked of other citizens? From a practical standpoint, it is virtually impossible to identify and measure the net impacts incurred by individuals from public intervention. Economists have offered the purist standard that intervention is justified if, as a result, some are made better off and none worse off, but by that standard few public actions would be possible. Where a net overall gain results, elaborate systems of compensation theoretically might be devised between gainers and losers to allocate the benefits satisfactorily. Or one might take the position that an action is justified simply by virtue of enlarging society's wealth, regardless of how that wealth is distributed. However, that position is very hard to reconcile with established property rights or notions of fairness. The inevitable recourse would be further complication of the tax code.

Finally, some reasons for intervening may have little to do with economic efficiency or rigorous equity. A nation's political and social stability rests on myriad interlocking bargains. It would be hard to say whether the federal support of home ownership, so often decried by economic purists, has been a net gain or loss if this question of social peace is included.

LAND AND FUTURE GENERATIONS

It was suggested earlier that unregulated private land will not adequately ensure the welfare of future generations in such cases as preserving unique or unusual ecological systems and genetic information, or preventing permanent environmental damages, because private owners may have insufficient incentive in each instance. Ecological systems on private land generally are too deformed to have unique value, but where such value occurs it rarely yields any current return to the owner. To deny an owner the right to put the land to other use would appear inequitable, and raises the question of why an individual should bear the disproportionate cost of meeting a social objective. Evidently public purchase of either all or part of the property rights is the only acceptable way (barring a gift) of ensuring the rights of the future.

While the cost of acquiring such rights is current opportunity cost, it is unclear how the long-term value of holding such rights is to be reckoned. An attempt to measure the value of preserving a species seems more than a bit of arrogance. Do we really have so much confidence in that intellectual construct of such recent vintage--neoclassical economics--that we would make it the arbiter of eons of evolution. To ask the question is not to pretend that there is a good alternative decision rule. But perhaps we should proceed rather tentatively in such matters. We face a dilemma created by our numbers and technological powers. It may be that every acre of the earth has some unique microbe, yet we cannot forswear possibly destructive use on some acres. Since the value of what is lost is unknown to us, attitudes of responsibility and respect for other life forms are needed for a decent compromise between immediate human economic demands and imprudent and irreversible destruction of potentially valuable genetic resources. Economic analysis brings only a very narrow perspective to this problem and need not pretend to govern the decision.

Preservation of the productive capacity of the land was identified as a long-term social objective. While land has the capacity to provide some forms of output in perpetuity, it may not be to the advantage of private owners to manage their land in that way. They may choose to treat it as a depletable resource. Farmers with erodible land face one stream of income if they manage their land so as to maintain the soil, and a different one if they manage it to maximize income. Even with a very low discount rate and a very long time horizon, they may find the destructive use economically rational. At some point continued erosion would reduce their soil's fertility and it would become rational to change the pattern of use. Calculations for fairly common conditions in rural America based on very modest discount rates and fairly severe rates of erosion suggest that private owners would not be motivated to reduce erosion for many decades and not before soil productivity is severely impaired. If more realistic current discount rates are used, there is even less incentive to practice conservation. And few farmers have the still longer time horizon that would be required at those rates (Seitz and coauthors, 1979).

This dilemma has frustrated soil conservation programs for decades. Farmers have been educated in good conservation practice. They understand it. They have adopted those practices that pay off in increased yields, and where erosion control is very low cost or is subsidized, they will employ it. But if good conservation practice requires more years in grass and fewer in corn, farmers often opt for income instead of conservation, and plant corn.

What will replace the missing topsoil? Farmers need not worry about that, since the question will be answered after their own individual lifetimes. However, society, which presumes to be at least as immortal as its capital, does worry about it. Without being quite explicitly stated, there is widely held sentiment that the capacity of the soil to produce should be maintained--that future generations should have access to land with substantially the same capacities as current land. This social standard does not discount future income from land as a private holder does. By implication, this means a willingness to forgo current income derived from soil depletion in favor of a level at which resource capacity can be maintained indefinitely.

Note that this willingness need not imply complete commitment to sustained use. Some productive uses are inherently unsustainable and may be treated as exhaustibles. For example, it might be impossible to maintain an erodible hillside in crops indefinitely at reasonable cost, but the hillside could still be cropped for a period before reverting to its next best use as pasture or forest, for which it might remain suitable.

This concern for future generations may be excessively cautious. Certainly our use of land can improve as well as degrade it. Improved drainage, leveling, better access, terracing, soil amendments, and other improvements all may permit land to be employed in more productive uses. Will future needs for land be as great? Technical progress that increases yields reduces the amount of land needed for a given output. That has been the dramatic story over the past few decades in agriculture, and it promises to be the case in forestry in the years ahead. Carried far enough, we might move to a system with very low requirements for land, soil, or natural fertility. However, the feasibility of such an approach remains to be demonstrated, and until it is, current resources cannot be safely dissipated. With continued population growth, perhaps future needs will be still greater.

If maintaining the productive capacity of land is accepted as a broad social objective, and if owners of private land do not find that to be in their interest, should the owners be required to support the social objective that they share with their fellow citizens? They may reasonably be asked to support it, but in equity there is no reason why the owner's share of the cost should be greater than that of nonowners. It is in the owner's interest, for example, to attain some level of erosion control, but beyond that level, the amount that society wants to buy should be made an explicit decision. Operational control of the land resides with farmers, and their cooperation is imperative, but it should be compensated. We have long had subsidies for soil conservation, but they have too often been distorted into income support or special interest purposes. If the public wants to buy a stake for future generations, that is what it should be getting, and it should be bought as economically as possible. Since farmers know their land best, they should be allowed to produce results in the most cost-effective manner, but they should be compensated by results--dollars per acre per ton of soil. With limited budgets, program managers would need to focus on the most productive of threatened soils.

Flexibility is also of value for the future. Conservation is one way of maintaining flexibility. The principal concern, however, is that too much land that is well suited to commodity production may be committed to urban use, from which it seldom is retrieved. The farmland preservation issue has elements of this concern, but unfortunately the issue is also so cluttered with other interests that it may be unfair to include it under concern about future generations. Those interested in farmland preservation include farmers or developers seeking tax-sheltered ways of holding land for speculation, local interests concerned about the immediate viability of their economic base, and urban dwellers seeking to maintain open space without public expense.

A long-range view of future needs for farmland or forest land is inevitably very murky. We cannot know remote future demand, and we do not know how far improved yields will go to meet demand with a given land base. Clearly, research to improve yields and policies to limit demand are alternatives to policies to retain land, although there are opportunities to retain land also. I would argue that such policies are more effective emanating from the city (aiming to shape and direct its growth) than from the country (striving to protect land). The economic power of urban uses to command land is simply too powerful to be contested in the countryside. The best that can be hoped for from land preservation plans is some control over quasi-urban use through agricultural zoning or districting (Brubaker, 1980).

PRIVATE GOODS, PUBLIC GOODS, AND EXTERNALITIES

Commodity output is the great strength of private lands, and most of the land best adapted to commodity output is privately owned. Private owners can manage farm and forest land sensitively and flexibily, emphasizing those outputs that yield the greatest net revenues without facing the same budget and other rigidities imposed on public land managers. However, certain structural problems, such as the matter of size, may prevent best use of private land. In post-World War II years, farmland was consolidated into fewer holdings of larger size, and once the human displacement was accomplished, the result was a far more profitable and progressive farm economy. No equivalent consolidation appears to have occurred on small forest holdings, in part because few owners have been dependent on them as a primary source of livelihood. As a result, the size of unit for farm woodlots and small forest holdings is not conducive to efficient management. Although they do fairly well, they do not produce in proportion to their total acreage and site class (Clawson, 1975). Better organization, perhaps through cooperative management arrangements, may help.

However, many small acreages of private forest land are not held for commodity production, but satisfy their owners' desire to own a piece of land or woods and perhaps carry an element of speculative interest. The same reasons apply to ownership of small tracts of farmland. Farmland held for development, speculation, or amenity purposes is not likely to be intensively farmed, and investments to ensure its continued productivity are apt to be neglected. In the countryside, the proliferation of small holdings that are primarily of residential, amenity, or speculative character often takes land out of farming or forestry and precludes its efficient management for those purposes. Is this a problem? In part it is a reflection of the relative abundance of land. In countries where land is in short supply, the smallest holdings may be the most intensively worked. However, the mixing of residential uses with active farming or forestry may have unhappy consequences for both. Pressures to abate the nuisances of commodity production grow as residential uses expand in the countryside, and the cost of public services to dispersed rural dwellings can be high.

There appears to be no acute reason for public intervention to alter this rural clutter. It is messy. It need not be encouraged by tax or other programs. But we do not presently want for commodities because of it. In the longer term, however, it could become a problem.

Most of the potential output of private lands is marketable--a much higher share than on public lands--and, in consequence, market signals are a more appropriate guide to use on private lands. Yet there are desired outputs from private lands--positive externalities and amenities and public goods--for which the owner is unable to collect and whose output is neglected. And, of course, there may be negative externalities that demand control.

Return to the question of erosion. Private owners concerned only with their own interests may allow erosion to proceed. Two kinds of damage results: the first involves the productive capacity of the land, as has been discussed; the second is off-the-field damage from sediment and entrained chemicals and nutrients and from increased flood levels. These are negative externalities inflicted on others by the private owner's operations. The law places the onus of control on the polluter, who may be required to contain this discharge. In practice, however, there is no effective public control of this problem, 208 water quality programs notwithstanding. An effective way to limit these damages is on-the-field erosion control, but there is no reason to require that method, however much we may wish to encourage it for other reasons. Less expensive methods such as border strips or runout areas may preserve water quality. Since the burden of control is placed on polluters, they are entitled to choose their own instruments. But alert public authorities should recognize that this is a bargaining chip for inducing farmers to use on-the-field erosion control measures which, at reduced public expense, will meet both purposes.

The control of negative externalities already has been made the responsibility of the polluter.[2] But what broader inducements might secure the optimum production of nonmarketed outputs from private lands? (These outputs include such services as watershed protection, habitat for wildlife, scenery, and accessible open space.) The usual assumption is that as long as competition prevails, markets operate so as to result in optimum output, but it is recognized that this does not operate in the case of public goods. The owner cannot profit. No attempt is made to identify actual or potential output of public goods on private lands. We may produce a disproportionate share of them on public lands at higher cost

---

[2]Few would shift the responsibility away from those who generate the problem. Most economists would argue for economic incentives that internalize environmental costs and allow them to be expressed in price as being more efficient than direct government controls. However, there is good legal basis in the policy power and in health and welfare provisions for direct controls.

because of sheet administrative and budgetary convenience. Private owners may provide some public goods as a by-product of other operations, but beyond that level, their optimum production would hinge on public purchase of the output.

Government intervention to secure such output would not be a simple matter, and in the case of low-valued public goods, the administrative or transactions costs of doing so might be prohibitive. Use of the police power could be considered--zoning for open space is an example--but this puts a burden of doubtful equity on the owner.

Some services of land are only quasi-public. Owners may be unable to profit because the values are too low to justify the expense of collection or of assuming the liabilities that go with offering a service--for instance, permission to hike, ski, or picnic on private land. Cooperative arrangements between groups of owners and user groups might be utilized to encourage such use. Government acquisition of easements is another possibility, though clumsy. Private holders may be able to provide some of the same services provided on public land at lower cost, but if the public services are unpriced, the owners' market is undercut and there is no incentive for them to do so. From a public viewpoint, there is no reason why the services should be provided at greater expense on public land, even though it may be infeasible for private operators to charge. A rational response would be to have public goods produced at lowest cost. Easements, leases, and service contracts all could be devices for more efficient provision of public goods from private lands, and if fuel shortages limit access to remote public lands in the future, such possibilities could become significant considerations (although we do not seem fated to have the same access to the countryside that a weekend walker would enjoy in Britain, where a different evolution occurred).

Much land, private or public, serves multiple uses, whatever its primary use. Recreational or amenity uses are a common form of secondary output. Public lands traditionally have provided such services without charge or at a nominal fee. The provision of such services by the public, however, is more a function of the capacity of public lands to produce them than of a conscious decision on the level of services to be offered and an analysis of where they are best provided.

PUBLIC CONTROLS AND PRIVATE PURPOSES

Most of the discussion so far has dealt with inherent market failure as a basis for intervention in land use decisions. Reasons for intervention include the protection of future generations against irreversible change and attempts to improve the efficiency of resource use in cases where the market does not operate effectively, as with the provision of public goods, efficient production of public services, or minimization of externalities. It is recognized that society has other goals that may be pursued through land use controls, whether or not that is the preferred course. Often these objectives can logically be pursued through public programs or incentives that relate closely to costs borne or services rendered.

The glaring exception to this is the control of urban land. Land taken into urban use is far more valuable than rural land. Within a developed area, in the absence of controls there would be disparities in value resulting from the desirability of certain locations, and there would be increases in value as density increased. With controls, these market factors are modified by the action of public authorities. On the urban fringe, the value at which land is taken into urban use is partly a function of the scarcity of land on which urban use is permitted, and hence that value is partly a result of the fact of regulation. There may be good public reasons for such intervention, as we have seen, but public action based on these reasons creates other problems.

Most urban land use conflict arises not over broad social objectives, for which there may be substantial agreement, but rather over how their implementation affects specific parcels. The matter is purely and simply a question of who gains and who loses, and the stakes are high. If the market for urban land operated freely, gains from development would be available to those who made accurate and timely judgments about the direction of growth. With controls, gains go to those with information about or influence over public action. Others may not only be denied gains, but may suffer losses as a consequence of undesirable nearby activities sanctioned by public action. If such results arise from influence over public action, they clearly have little objective defense and may be corrupt. What is the government doing in this game, making some rich and others poor by controlling a market for land? If public action creates value, that value does not logically become the right of individuals to claim, however greedily we all look for the chance to do so.

Because the stakes are high, it is all the more imperative that the social rationale for intervention in land use markets be clear, and, that, in the absence of a clear rationale, it be minimized. The fact that land is property carries implications both under the law and in terms of equity. Government must not be a party to private purposes. Much intervention has become routine and has lost contact with its social rationale.

The socialization of development gains arising from public action would eliminate much of the incentive that now so corrupts the system, and it might also mean that land would shift into urban use at lower cost. One proposal for accomplishing socialization of development gains, associated with Clawson and others, would be to auction zoning rights. Under this proposal, whatever public interest there is in maintaining the control and compactness necessary for efficient services could be preserved while at the same time much of the gain from public action could be captured. Planning could still designate areas suitable for development, provided there were enough alternatives to permit competition for rights. There is no reason why this approach should not be as efficient as any other regulated approach for granting permission to develop.

Another approach, transferable development rights, leaves development gains in private hands but spreads them over a larger group of recipients and thus lowers the stakes. In this system, each owner is accorded development rights in proportion to his or her land holdings, and these rights may be marketed apart from the land. Anyone wanting to build at greater density than allowed by the land's original rights would need to buy rights from others who would surrender the right to develop. Planners could still designate areas to which rights could be transferred and thereby maintain some control over the direction of development as well as of total density. Disputes over gains and losses should be eliminated through this technique, which is a compromise between the widely held view that gains should remain in private hands and the reluctance to allow arbitrary government action to determine who gets them. So far, it is little tested.

Despite such proposals, most development is governed by zoning actions that create artificial shortages, leave gains in private hands, and provide maximum incentive to influence the system. Zoning may also be used to defend neighborhood quality, a result that most view with favor even though it may delay a shift to more intensive activity that would make best use of the land. There is little reason to suppose that zoning results in the most efficient use of land. It can, however, help minimize exposure to externalities.

CONCLUSION

Significant public purposes can be sought through land use controls. The market performs most of the allocative functions in regard to land, but it does not serve all efficiency functions, and it takes little account of very long term resource objectives or of current social objectives. The main difficulty with intervention involves the equity problems it creates, even in the best of circumstances. These problems become overwhelming when the effort involves regulating high-stakes games such as urban land markets in which the fact of intervention raises the stakes. To the extent that regulation is unavoidable, it might have greater chance of success if the amount of windfall gains riding on outcomes were reduced.

REFERENCES

Brubaker, Sterling. 1980. "Agricultural Land: Policy Issues and Alternatives." Paper presented at Resources for the Future conference, Adequacy of Agricultural Land, Washington, D.C.

Clawson, Marion. 1975. Forests for Whom and for What? (Baltimore, Johns Hopkins University Press for Resources for the Future).

Erwin, David E., James B. Fitch, R. Kenneth Godwin, W. Bruce Sheperd, and Herbert H. Stoevener. 1977. Land Use Controls (Cambridge, Mass., Ballinger).

Seitz, Wesley D., C. Robert Taylor, Robert G. R. Spitze, Craig Osteen, and Mark C. Nelson. 1979. "Economic Impacts of Soil Erosion Control," Land Economics (February).

U.S. Department of Agriculture. 1973. Major Uses of Land in the United States: Summary for 1969. Agricultural Economic Report No. 247 (Washington, D.C.).

_____. 1975. Perspectives on Prime Lands (Washington, D.C.).

Richard Stroup

COMMENTS

My comments on chapter 3 will briefly reiterate the important points made by Sterling Brubaker and will then move on to some disagreements I have regarding his statements of fact. Most of my comments concern what I interpret to be ideas that are commonly held, but which reflect a misunderstanding or incomplete application of the fundamental notions of political economy.

Brubaker makes several valuable observations regarding the operation of the public sector in the area of land use management. For example, regarding our obligations to the future he notes that "our political instruments for expressing this interest are disastrously short-shighted." Decisions that inflict pain now in order to gain benefits for unborn generations are certainly not very popular with representatives who face election contests every two, four, or six years. Congressional action in the areas of inflation and the Social Security program over the past twenty years drive home this point with a vengeance.[1] The extent to which we can logically entrust our natural resource patrimony to this decision-making process is at best questionable. But what is the alternative? I will return to this salient point shortly.

Brubaker points out that a good deal of the urban sprawl problem is caused by low-priced public services. The shortsighted fixation of the political process on keeping service prices low (generally at average instead of marginal cost) has led to an implicit subsidy of socially in-

---

[1] For the reasoning on how and why, even with well-intentioned political actors, the "shortsightedness effect" is powerful in a democracy, see Gwartney and Stroup, 1980, chap. 32.

efficient developments.[2] This point is not new, of course, but I believe its importance is growing substantially—in the field of energy conservation as well as in the field of land use management.

Brubaker's third point—commonly overlooked by many commentators on land use policy—is that those who influence public policy on land management have mixed motives. While some come to that arena out of a strong sense of public spiritedness, a great many cloak their self-interest in the rhetoric of "the public interest." It is a simple though not always understood fact of political life that actions taken in the name of public policy nearly always benefit incumbent political forces. Whether the issue is rent control, building codes, or land use concepts, it is nearly always true that those currently in possession of whatever is being regulated and currently active in the political jurisdiction are protected at the expense of those who may arrive later and wish to acquire access.

Another point Brubaker makes is that public goods associated with land management can be provided not only by public ownership of the land but also through the purchase of easements by the public sector. As an economist, I am always delighted when public policy tools can make the costs or the benefits of an action explicit. The purchase of easements does exactly that. It also has the advantage that productive capabilities not needed for the public good provision are not purchased. Thus, the cost of providing the public good is minimized.

Regarding the two statements of fact with which I must disagree, the first is that "land in virtually any urban or infrastructure use (including housing) is more valuable than the same land used for agriculture or forestry. As a consequence, urban use makes first claim on land whenever the market governs." This simply is not true. On more than 90 percent of all private land, there is no urban development. The very thought that urban uses are superior on most of the private land in Montana east of Billings is guaranteed to bring a grin to the face of any Montanan. Brubaker, of course, knows that this applies only to already urbanized areas, and yet seemingly semantic slips can be important. This type of

---

[2] The mechanics of how low public utility prices (generally based on average, not marginal cost) lead to excessive dispersion of development, or "sprawl," are explained in Randall, 1981, chap. 19.

lexicographical ordering in which certain <u>categories</u> of use for a resource are said to be more valuable than other categories ignores the marginal principle of economics. At any point in time, certain land will have extremely high value in urban use. However, as one moves farther away from urban centers, the value of bringing more land into urban use declines. For any parcel of land, the issue is what the marginal costs and benefits of changing its use are.

A second statement at odds with the facts is that "owners of property are entitled to its exclusive use." Some years ago this may have been close to the truth. In recent decades, however, landowners have become much less free to do as they wish and more beholden to outside decision makers. Increasingly, government has used police powers to limit the owner's use in order to preserve benefits accruing to persons not owning the land. The extent to which owners really are entitled to exclusive use is small and is decreasing (Johnson, 1975; Johnson, chap. 5, this volume).

Moving from questions of fact to the applications of economic analysis, I want to deal with two fundamental issues. The first is the role of economics, and the second is the way in which the market, as contrasted with the government, handles resource allocation over time. Brubaker underestimates both the usefulness of economics and the future-oriented nature of private property ownership.

Brubaker says that "economic analysis brings only a very narrow perspective to this problem [the consideration of the welfare of future generations]." This statement may well apply to any particular <u>economist</u>, but the applicability of economic <u>principles</u> is just as valid here as it is in analyzing business decisions. Positive economics, as the science of choice, is not wedded to a particular objective function, such as profit maximization. Indeed, the marginal principle, the first law of demand, the notion of opportunity cost, and the realization that exchange is not a zero-sum game are principles that apply not only to the world of business but also in the long run to the world of resource conservation. These principles are just as useful to a saint struggling to save souls under a limited resource constraint as they are to a business administrator practicing his trade.

This observation is not trivial. Recognition of the marginal principle, for example, causes us to question lexicographical orderings of the type that state that certain land uses are preferable to certain other land uses. The more relevant question, of course, is the extent to which <u>more</u> of a particular land use has value relative to more of a <u>different</u> land use. Recognition of the concept of opportunity cost allows one to deal rationally with the emotional cry that economists always reduce resource values to dollar terms or that they think in terms of human value versus dollars. The real choice is always between opportunities. Money prices, of course, frequently carry strongly validated information on how market participants value these trade-offs at the margin. Brubaker vastly underestimates the importance of economic principles. So long as we are dealing with economic principles, as opposed to complex and specific economic models, the applicability is extraordinarily broad.

Resource economists who discuss market management of natural resources versus governmental control generally have a pretty good understanding of how markets work. However, the resourcefulness of profit-seeking entrepreneurs is continually underrated. For example, probably every resource class in the country covers neighborhood, third-party, or external effects. Far fewer of those same classes will show how developers of shopping centers, residential areas, or golf course/residential developments organize their projects so that most of the positive externalities are captured in land values while negative externalities are reduced by means of protective covenants.

Nearly every form of governmental regulation is the <u>creation</u> of third-party effects. After all, regulations are normally the imposition by a third party (government) of conditions permitted to exist between consumers and owners/producers. To what extent is a Cecil Andrus or a James Watt held personally responsible for what he or his employees do in the course of managing Bureau of Land Management lands? At the core of each market failure is the fact that a decision maker is not held fully responsible for what he or she does. Social payoffs, therefore, differ from personal payoffs. Can anyone name a government official for whom that condition does not hold regarding virtually every decision that official makes?

Market failures clearly do exist, and many of them are important. Nearly all of these stem from the fact that easily enforced and transferable property rights to the resource in question either are held in common or do not exist. Grizzly bears and blue whales are in danger precisely because, unlike exotic animals on Texas game ranches, they are unowned. To take this as a signal to make all property rights automatically communal is simply absurd. Just as we cannot assume the existence of a perfectly competitive market (without externalities) for natural resources, we cannot assume the availability of a benevolent, omniscient, and omnipotent governmental agency standing by to straighten out the slightest market failure. Every public decision is a public good. Too little decision effort (and too much lobbying) is typically applied. The choice is always among imperfect alternatives.

Probably the gravest error committed by resource economists in general, and Brubaker in particular, is the assumption that shortsightedness will dominate in market situations, relative to government. A strong logical argument can be made that precisely the reverse is true. As mentioned earlier, Brubaker does recognize the shortsightedness of current governmental institutions. Beyond that point he seems to forget that fact and does not ever seem to realize the inevitable lack of any signal or incentive for governmental decision makers to provide for future generations. In the market, one's charitable instincts with regard to the future are reinforced by market incentives and signals. Not so in government, where only charitable instincts help future generations.

If we assume that individuals tend to be shortsighted, since each has a finite life and is at least partially self-interested, then for a resource whose benefits can be packaged and sold, the case for this view is strong indeed (Stroup, 1977; Stroup and Baden, 1979). The argument revolves around two characteristics of markets. First, the asset price of the resource is the voice of future demanders of that resource, speaking to the present. Second, it is not the holder of the median opinion that governs use of the resource, but rather the high bidder.

Who will own a piece of land? The high bidder (or the owner who rejects the high bid) values the property most. That is, the owner will be the person or group that places the highest estimate on what the present discounted value of all future services from that land will be. The determination of that value, of course, is complex. It will depend on future demand for those services, the price and availability of substitutes and complements in production, and the time rate of discount. Those who think the future value will be high or who have a low discount rate will tend to prevail in the market. In the absence of a smoothly functioning futures market for land, those with medium or low estimates of the land's future value, since they cannot sell land short, will simply have no influence in this market. This contrasts rather strongly with a democratic setting in which the median voter opinion, for example, as to whether a piece of land should be developed, will presumably prevail.

Why does the high-bidder-wins property of the market yield a bias in favor of the future, relative to median voter control? Consider the case of a copper mine that can be mined today or saved for mining at some future time. Assume that, on average, people believe that the payoff for waiting will be the same as the value of mining now: that is, the median opinion is that the two payoffs are similar. Compared with the estimates of what current exploitation is worth, we could expect a greater divergence between those at the high end and those at the low end of the expected value spectrum in the case of estimating the future value. Uncertainty about the future creates a widespread distribution of estimates, even though the mean is assumed to be the same. If the decision is made on the basis of votes, with the majority ruling, then the vote would be even. But if the decision is left to the market, where everyone knows about what the value in _current_ exploitation is, nearly half the population would be interested in speculating and holding the copper ore off the market. The cynic may ask, "But what about the people who think copper will be valuable later but who do not expect to live long enough to see the copper mined?" No problem. So long as the asset is transferable, any speculator can gain by watching the value rise, regardless of how many centuries it will be before the optimal extraction time has arrived. So long as the value of the asset is expected to rise in real terms at least as fast as the _real_ interest rate, then holding the copper _in situ_ is a good private and social investment. Of course, if the resource price fails to grow that fast, then holding the ore off the market would have been a poor investment, both privately and socially. After all, the real interest rate measures a social as well as a private opportunity cost. The cynic might again ask, "But what if people have a lower discount rate for social decisions than they do for private ones." To answer, I ask why individuals should be more eager to bequeath a gift to society in general than they are to leave something to their own heirs or assignees. If I am to leave a gift for the future, I would prefer that it be passed down through generations of my own family.

An inescapable fact of life is that the current crop of decision makers will decide how much of our natural resource patrimony (and alternative items of value) will be consumed now and how much will be left to the future. This much is commonly recognized. What Brubaker and others miss is the extent to which profits reward those private decision makers who recognize future resource scarcities and provide for them through speculation. Each speculator captures the value of waiting by selling the asset he or she holds--perhaps to another who will carry on the speculative activity. When property rights to land or any other asset are privately owned, that asset is more likely to be saved by speculators than if a median opinion decided the issue. (I assume here that mixed portfolios can neutralize risk aversion for sufficient speculators.) Without a futures market in the resource, too much of the resource will probably be preserved. Of course, losses will penalize the unsuccessful speculators.

The market's excessive future orientation regarding exhaustible resources is matched by its pressure for good stewardship on renewable resources, such as timberland or farmland. The asset value is truly the voice of the future speaking through the high bidder. Farmers who allow erosion to reduce the future productivity of their land see their wealth reduced by the present value of that future stream of income. The land value falls as soon as an appraiser can see the difference. The decline in land value may in fact be tolerable if large enough advantages accrue to offset the future losses into perpetuity. Once again, however, note that potential bidders with low discount rates and high opinions of farmland's future value determine the relevant discounted present value that is the farm's value today. Not only is farmland's price the voice of the future speaking to us now, but the landowner's wealth is hostage to good land stewardship. Anyone who allows the future productivity of his or her land to decline suffers the entire decline now (overstated by low interest rates and optimistic net crop value estimates).

How do these market incentives, which reinforce any chartiable instincts current decision makers have toward preserving natural resources for the future, compare with governmental incentives? Since future generations cannot vote and since governmental managers cannot sell the appreciated value they might build into the land by investment and the sacrifice of current use, future generations depend entirely on the charitable instincts of current voters. In government there is no analogue to asset price or speculation as the voice of the future. Speculators are cursed for holding resources off the market; profits allow them succor. Being cursed by current voters is not a survival trait in politics.

Good intentions are not enough. Information and incentives, which are largely determined institutionally, are crucial. Let me finish these comments by quoting Arthur Young, Britain's agricultural minister in 1787: "Give a man the secure possession of a bleak rock, and he will turn it into a garden; give him a nine years lease on a garden, and he will convert it into a desert." Either this is an exaggeration or we are very lucky that each federal administration arrives with only a four-year lease.

REFERENCES

Gwartney, James, and Richard Stroup. 1980. *Economics: Private and Public Choice* (New York, Academic Press).

Johnson, M. Bruce. 1975. "Land Use Planning and Control by the Federal Government," in *No Land Is an Island* (San Francisco, Institute for Contemporary Studies).

Randall, Alan. 1981. *Resource Economics: An Economic Approach to Natural Resource and Environmental Policy* (Columbus, Ohio, Grid Publishing).

Stroup, Richard. 1977. "Property Rights and Natural Resource Exploitation." Staff Paper (Bozeman, Montana State University, Department of Agricultural Economics and Economics).

_____, and John Baden. 1979. "Property Rights and Natural Resource Management," *Literature of Liberty* vol. 2, no. 4, pp. 5-44.

Chapter 4

STATE INTERVENTION ON PRIVATE FORESTS IN CALIFORNIA

by Henry J. Vaux

Public intervention in the use and management of private property has had a long history in California. In 1852, little more than a year after the state government of California was established, the state assembly adopted a resolution that reads in part as follows:

> WHEREAS the lands of California are sparsely wooded and timber for building purposes is extremely scarce and difficult to be procured...
>
> BE IT RESOLVED that our Senators in Congress be instructed and our representatives requested to . . . procure the passage of a law whereby the settlement and occupation of all public lands upon which Redwood is growing shall be prohibited and the Redwood timber shall be declared to be the common property of the citizens of California for their private use and benefit, provided such timber shall not be made the subject of trade and traffic (Clar, 1959).

Unfortunately, the resolution did not make clear how the banning of all trade or traffic in redwood could alleviate the scarcity of timber for building purposes that the assemblymen recognized. But considering this early resolution, it is not surprising to find either that the California legislature has, over the ensuing 130 years, intervened in private forest management affairs in a variety of ways, or that these interventions collectively have still not fully quieted active political debate over how to achieve both a socially adequate supply of timber and a socially acceptable degree of environmental protection.

This chapter consists of three parts. First, it describes the present system of state intervention on privately owned forest lands in California, including brief reference to the historical and cultural circumstances that gave rise to the system. Second, it analyzes how each such intervention may have influenced current and future levels of output from private forest lands and considers the related matter of the effect of intervention on incentives for investment. Finally, it provides an evaluation of the cumulative effect of the various types of intervention on forest management, land use decisions, and the achievement of the state's statutory goals in forest policy.

THE CALIFORNIA SYSTEM OF INTERVENTION
The California system of timber harvest practice regulation cannot be understood unless it is placed in the broader contexts of the state's particular social and physical environment and of the entire forestry policy of the state, both of which are discussed in the sections that follow.

Physical and Social Environment
Because of California's Mediterranean type of climate, its people were actively and politically concerned with questions of wildfire, flood prevention, and water quality very early in the state's development. In response to these concerns, the California Board of Forestry was established by the legislature on March 3, 1885 (Clar, 1959). The state has been continuously involved in forest policy activity affecting private landowners more or less throughout the following ninety-five years. The concerns over wildfire control, watershed protection, and water quality remain as fundamental to policy today as they were then.

Another characteristic of the state that has been a strong and continuing influence on forest policy is the fact that most Californians perceive the forest in terms of urban value systems. Since before 1900 a majority of the population has lived in urban places, and today more than 90 percent of the population is urban. Between 1940 and 1980, the number of urban residents grew by almost 16 million.

Urban perceptions of forest values were expressed in California at a relatively early date. Three national parks were established in the state in 1890 with strong support from the citizenry, and in the same year the Sierra Club was incorporated to marshal support for wilderness values. California's citizens invested major effort and direct financial aid in the establishment of a unique system of redwood state parks beginning in 1920. More recently, 1.6 million gross forested acres have been set aside as wilderness area, with similar action on an additional 2.0 million to 4.7 million gross forested acres currently pending before the Congress or other decision-making body (Calif. Dept. of Forestry, 1979a).

In comparison with the situation in most timbered states, the process of converting California's virgin forest to forest management came late. Sawmilling in both the redwood and Sierra Nevada regions began before the gold rush, but as recently as 1953, three-fourths of the public land and over half of the private land in the state were still dominated by old-growth timber (USDA, Forest Service, 1954). More than 75 percent of all the timber ever cut in the state has been harvested since 1940. Given the mobility of the California population and its predilection for recreation in forested areas, the liquidation of the virgin forest there has taken place in full view of a large segment of the urban body politic. The impact of these events on voter perceptions of forests and forestry should not be underestimated (Vaux, 1980).

SCOPE OF CALIFORNIA'S FOREST POLICY

From the geographic and political origins outlined above has come a broad state forest policy providing for various forms of intervention on private forest land. The policy consists of five major elements: (1) protection of private forests from fire, insects, and disease; (2) provisions for forms of taxation of forest property and income that recognize the peculiar financial requirements of forest management; (3) financial aids to private owners to encourage rehabilitation of forest land as needed to achieve reasonably full site capacity; (4) research and professional education programs; and (5) careful regulation of timber-harvesting practices aimed at ensuring continued forest productivity and protection of environmental values. The policy applies to some 7.6 million acres of privately owned commercial forest land in California, as shown in table 4-1. This is 47 percent of the current commercial forest area of the state. Most of the remaining commercial forest is in the National Forest System. Of the privately owned land, 35 percent is owned by forest products corporations, 18 percent by large corporations that do not manufacture forest products,

Table 4-1. Commercial and Noncommercial Forest Land in California, by Class of Ownership, 1978

(thousand acres)

Ownership class	Commercial forest land	Productive forest land reserves and forest land with less than 20 ft^3/yr productivity	Total
Industry:			
Forest products	2,688	n.a.	n.a.
Non-forest-products	1,388	n.a.	n.a.
Nonindustrial owners	3,556	n.a.	n.a.
All private	7,632	5,531	13,163
Public agencies	8,667[a]	10,728	19,395
Total (all owners)	16,299[a]	16,259	32,558

Source: Calif. Dept. of Forestry, 1979a, table 3-1.

Note: n.a. = not available.

[a] Carter administration proposals for the RARE II (Roadless Area Review) area would have reduced this area by approximately 635,000 acres.

and 46 percent by nonindustrial private owners, most of them with small holdings. The state's forest policies have been designed with some attention to this diversity of private ownership.

Basic fire protection on private forest and other watershed lands is provided at the expense of the state's general fund. This resulted in an annual outlay of general fund monies under the 1979-81 California budget acts averaging $87.9 million per year for fiscal years 1979 to 1981. The prorated portion of this sum, chargeable to protection of private commercial forest land, is $23.9 million per year. The five-year average burn in the major commercial forest land districts is 0.39 percent. Thus, state policy provides a level of protection from fire adequate to permit rational long-term management on the private forest land protected.

The state's taxation policies have generally been favorable to forestry since 1926 when essentially all timber under forty years of age was exempted from the ad valorem property tax. In 1976, with passage of the Forest Taxation Reform Act, and in subsequent years, the tax system was further modified to provide for (1) a yield tax in lieu of an annual ad valorem tax on all timber, and (2) land classified in timberland preserve zone to be taxed on the basis of value for timber growing rather than on value in "highest and best use." Two years later, additional legislation--the Forest Improvement Act of 1978--provided (3) quick amortization of planting costs as a deduction from state-taxable income, and (4) modification of state inheritance taxes on small nonindustrial forests to reduce the pressure to liquidate the forest estates of decedents.

Beginning July 1, 1980, the state initiated a program of grants-in-aid to small nonindustrial forest owners to encourage reforestation, stand improvement, wildlife habitat improvement, and other measures needed to bring depleted forest areas into full production. The program, authorized by the Forest Improvement Act of 1978, may subsidize up to 90 percent of the costs of such measures. It is funded by receipts from the sale of timber on state-owned forests, which previously went into the general fund. Appropriations for this purpose during the current fiscal year (1981-82) are $4.5 million. This investment by the state is backed up by more traditional programs of providing low-cost planting stock, technical assistance to owners, and extension forestry services at substantial state expense.

Forestry education and research have been carried on in California since 1914. The programs have been generally of high quality, although in recent years the scale of the research effort has clearly fallen behind expanding needs (Calif. Board of Forestry, 1979). As with the other elements of the state's forest policy, education and research have been supported to date very largely from general fund appropriations.

The four types of state interventions introduced above and discussed at greater length later in this chapter all serve to encourage private forest management. As shown in table 4-2, in the aggregate they represented an outlay of about $34.6 million per year of state funds, the equivalent of about $5.30 per acre of eligible privately owned commercial

Table 4-2. State of California Budget Provisions for Intervention in Private Commercial Forest Resource Management: FY 1979-81 Average

Function	Appropriation (thousands)	Percent of total	Outlay per acre of eligible land
Protection of forests:			
From fire (state funds only)	$23,877	65.2	$3.13
From pests	320	0.9	0.04
Aids to small owners			
Forest improvement subsidy[a]	4,497	12.3	1.26[b]
Service forestry	639	1.7	0.16[b]
State nurseries	544	1.5	0.07
Education, research, analysis, and state forests (includes University of California)	3,803	10.4	0.50
Regulation of forest practices	2,867	7.8	0.38
Foresters' registration	69	0.2	0.01
Total appropriations	$36,616	100.0	$5.57
Less: Applicable reimbursement and federal funds	2,043	5.6	0.27
Net total state funds	$34,573	94.4	$5.30

Source: California Legislature, Budget, 1980-81 (Sacramento, 1980).

[a] For 1980-81, the first full year of program.

[b] Applicable to approximately 3.556 million acres of small nonindustrial forest land only.

forest land each year. These aspects of the policy should be kept in mind with respect to the evaluation of the impacts of forest practice regulation later in this chapter, since a state that each year invests some $3.85 to $5.30 per acre of privately owned forest land in support of timber growing has a perfectly appropriate concern that public investment shall be reasonably protected.

Regulation of Timber Harvesting

State regulation of timber harvesting on private land began in California with adoption of the Forest Practice Act in 1945. Principal support for that legislation came from the forest products industry and forest landowners. Forest practice rules were developed by statutory District Forest Practice Committees. After approval by 75 percent or more of the landowners concerned, the rules were approved by the Board of Forestry, at which point the rules had the force of law. A majority of the members of the District Forest Practice Committees were, by statute, to be drawn from the forest industry and land ownership groups.

This system functioned from 1946 until late 1971. At that time the authorizing statute was declared unconstitutional on the ground that delegation of the rule-making authority to district committees whose majority represented the interests being regulated was not proper (Bayside Timber Co. v. Board of Supervisors of San Mateo County, 1971; Arvola, 1976).

This first regulatory system required a permit to log and provided that logging be conducted in accordance with the forest practice rules adopted by the Board of Forestry. Criminal penalties were ultimately provided for logging in violation of the rules, and some violators were prosecuted.

The twenty-six year period during which this regulatory act was in force covered the peak years of private timber cutting in California. When the act became law in 1946, the annual cut from private land was about 400 million cubic feet (ft^3). In 1953 it had risen to almost 900 million ft^3. By 1970 it had declined to about 450 million ft^3, the level at which it has currently roughly stabilized (table 4-3).

During its early years, the first Forest Practice Act was regarded largely as an educational tool (Nelson, 1952). In such terms, it had some

Table 4-3. Timber Removal in California by Ownership, 1947-78 (million ft^3)

Year	Total	Public	Private
1947	531	n.a.	n.a.
1948	664	n.a.	n.a.
1949	633	n.a.	n.a.
1950	717	n.a.	n.a.
1951	831	n.a.	n.a.
1952	843	110	733
1953	998	110	888
1954	933	131	802
1955	1,003	181	822
1956	977	195	782
1957	891	169	722
1958	946	199	747
1959	981	265	616
1960	856	240	616
1961	890	249	641
1962	923	249	674
1963	910	291	619
1964	913	329	584
1965	879	343	536
1966	835	342	493
1967	844	338	506
1968	891	419	472
1969	834	350	484
1970	761	327	434
1971	798	367	431
1972	834	392	442
1973	821	353	468
1974	784	306	478
1975	722	267	455
1976	789	331	458
1977	792	301	491
1978	777	314	463

Sources: California Department of Forestry, Production of California Timber Operation in 1977, State Forest Note 74 (Sacramento, Calif., 1979). USDA, Production Prices, Employment and Trade in Northwest Forest Industries (Portland, Ore., Pacific Northwest Forest and Range Experiment Station, 1980).

Note: n.a. = not available.

significant positive effects. Enforcement was strengthened during the
1950s, but by the mid-1960s there were increasing complaints of inadequate
environmental protection (Calif. Legislature, 1966, 1972). Thus, by the
time of the court action in 1971, there was already substantial political
controversy over the adequacy of the regulatory system.

In the months following the determination that the Forest Practice Act
of 1945 was unconstitutional, the industry generally followed the previously existing forest practice rules on a voluntary basis. Two or three urban
counties already had logging ordinances in place that regulated timber
cutting in those areas, but elsewhere there was no regulation during most
of 1972 and 1973. The legislature used this period first of all to commission a study of the problem, which resulted in the recommendation of
comprehensive legislation (University of California, Davis, 1972), and then
to draft the Z'berg-Nejedly Forest Practice Act. That act was signed by
then-Governor Ronald Reagan and became law on January 1, 1974. Thus,
California recently completed the seventh year of operation under the new
system.

Z'berg-Nejedly Forest Practice Act

The Z'berg-Nejedly Forest Practice Act of 1973 was designed to remedy the
constitutional defects in the earlier regulatory scheme, to provide broader
protection for environmental values than had existed previously, and to
install a more rigorous system of planning for and enforcement of the
necessary forest practices.

To achieve the constitutional objective, the new law reorganized the
Board of Forestry and the committees--now called District Technical Advisory Committees--and placed the rule-making authority clearly with the
board. Membership of the board and the committees was changed so that
five members of each body (a majority) are persons with no significant
ownership or financial interest in the industry. These so-called public
members must have some experience or background relevant to forest management and planning problems; these qualifications are broadly drawn and
interpreted. Three members must be appointed from the forest landowner/
forest industry sector and one member from the range/livestock industry
sector. All members are appointed by the governor, confirmed by the
senate, and serve four-year staggered terms. One of the public members

is appointed by the governor as chairman and serves in that capacity at the pleasure of the governor.

The legislature's intent to provide broader protection of environmental values is expressed in sections 4512 and 4513 of the Public Resources Code, which read in part as follows:

> 4512 (c). The Legislature thus declares that it is the policy of this state to encourage prudent and responsible forest resource management calculated to serve the public's need for timber and other forest products, while giving consideration to the public's need for watershed protection, fisheries and wildlife, and recreational opportunities alike in this and future generations.

> 4513. It is the intent of the Legislature to create and maintain an effective and comprehensive system of regulation and use of all timberlands so as to assure that:

> (a) Where feasible, the productivity of timberlands is restored, enhanced, and maintained.

> (b) The goal of maximum sustained production of high-quality timber products is achieved while giving consideration to recreation, watershed, wildlife, range and forage, fisheries, and aethestic enjoyment.

Although these statements of intent make it clear that the purpose of the California regulatory scheme is <u>both</u> to provide for maximum sustained production of timber and to provide protection for forest-derived water, fishing, wildlife, and recreation values, the relative weights to be assigned to these several values are deliberately left vague. Nevertheless, standards to be achieved in reforestation are made explicit in the statute. There is a specified minimum standard of stocking, either in terms of number of trees or of basal area per acre, which must be met within five years after timber harvesting. A higher standard may be set by the board, but the law establishes a floor below which stocking will be regarded as unsatisfactory. If stocking standards are not met within five years, the law requires that artificial regeneration be undertaken at the landowner's expense to achieve at least the statutory standard.

<u>Other Applicable Statutes</u>

Since adoption of the Z'berg-Nejedly act, the court has held (<u>NRDC</u> v. <u>Arcata</u>, 1975) that the principles (although not all of the procedures) embodied in the California Environmental Quality Act of 1970 apply to forest practice regulation, and that, where statutes appear to overlap in their jurisdiction, the Forest Practice Act must be construed, if possible, in ways consistent with other related state and federal legislation. The Board of Forestry interprets this to mean that maintenance of a permanent supply of forest products and the protection of water quality in conformity with standards established under the Porter-Cologne act of 1969 (California's basic water quality control law) are of coequal importance.

The weight that the Z'berg-Nejedly act gives to resource and environmental values other than water is of somewhat lesser magnitude when compared with that given to timber growing and water quality. The Board of Forestry in effect interprets what is "due consideration" for these other values. As a result of section 208 of the Federal Water Pollution Control Act amendments of 1972, the board is presently reviewing and to some extent revising the forest practice rules to ensure that equal weight is given to timber production and water quality protection (Calif. Board of Forestry, 1980c).

Other legislation must be recognized in applying the Forest Practice Act of 1973. Provisions of the state's Species Preservation Act and Endangered Species Act of 1970 may need to be reflected in Forest Practice Act regulations affecting habitat for certain plant and wildlife species; California's Coastal Act of 1976 requires special regulations for timber harvesting in special treatment areas designated by the Coastal Commission; and its Wild and Scenic Rivers Act of 1972 also may lead to special forest practice requirements in specifically designated areas.

These several statutes result in a distribution of authority among a number of state agencies. Under California's system of government, there is a considerable number of so-called independent boards and commissions (for example, the Fish and Game Commission, Water Resources Control Board, Parks and Recreation Commission, Mining and Geology Board) each having certain "lead agency" responsibilities with respect to resources, including forest resources. Within its area of responsibility, each of these boards and commissions may exercise control over forest operations. In order to minimize the adverse impact of this divided authority on those who are regulated, a conscious effort is made to achieve "one-stop shopping," under which the requirements of the several state agencies can be handled through a single regulatory process.

Preparation and Review of Timber Harvest Plans

To provide rigorous standards for planning for and enforcing appropriate forest practices, the Forest Practice Act provides that no timber shall be cut except in accordance with an approved timber harvest plan (THP). In effect, a THP must be prepared for each separate logging operation. A single plan may encompass as few as 3 acres or as many as 1,000 or more. The average area covered by plans approved in 1979 was about 245 acres.

Timber harvest plans must be prepared and signed by a registered professional forester (RPF) licensed under the Professional Foresters Licensing Act of 1972. In general, the plan must describe the silvicultural methods to be used, indicate the methods to be used to regenerate the area after cutting in order to meet the standard of restocking established by law, outline the methods to be employed to avoid excessive erosion, specify provisions needed to protect "unique" areas within the area of timber operations, and where necessary indicate how the forest practice rules will be complied with in the specific conditions of the plan area. There is provision for notifying the public of the filing of a THP; such plans must be available for public inspection; and they must be transmitted to the state Department of Fish and Game and the appropriate Regional Water Quality Control Board and County Planning Agency. (Independent of the THP process, the Department of Fish and Game must approve each stream crossing that is to be constructed, and the Regional Water Quality Control Board may establish waste discharge requirements for the operation.)

The director of the Department of Forestry, with whom the THP is filed, has up to twenty-five days to approve or disapprove the plan. During the interim the director may require a preharvest inspection of the plan area. About three-fourths of the plans have such preharvest inspections. Representatives of the Department of Fish and Game, Water Quality Control Board, and other concerned agencies may participate in the preharvest inspection. In each case, before taking action, the director requires consideration of the THP by an interdisciplinary review team. The fish and game and the water quality agencies are requested to provide representatives on the review team, and they may file dissents from the review team recommendation.

In the light of review team recommendations, the director approves or denies the THP. Denial must be based on a determination that the plan does not conform to one or more of the rules of the board. No other criterion may be used. In the event of a denial, the applicant has ten days to appeal the decision to the Board of Forestry. The board must hear the appeal within thirty days. The only appeal from a director's decision to approve a THP is to the courts. Upon approval of the THP logging operations may begin. The approved plan has a life of three years, which is not extendable. Operations under the THP must be conducted by a licensed timber operator (LTO), whose license is issued by the Department of Forestry and may be suspended or revoked for cause.

The LTO must conduct the operation on the ground in full conformity with the provisions of the THP and with the forest practice rules. The department is authorized to go on the ground to inspect the operation for compliance with the plan and the rules.

Enforcement Provisions

In a year, the department makes approximately 7,500 THP inspections, an average of about one inspection per plan in effect. In practice, inspections tend to be focused on those plans where problem situations are most likely to arise, so some plans receive considerably more than the average number of inspections (Calif. Dept. of Forestry, 1977-79).

Violation of the Forest Practice Act or of the forest practice rules is a misdemeanor punishable by fine or imprisonment or both. In addition, the department may bring court action to enjoin a violation or threatened violation. Where a violation has occurred, and upon a court finding that immediate and irreparable harm is threatened to soil resources or to the waters of the state, the violator may be ordered to take corrective action or the department may take such action, with any expenses incurred through such action becoming a lien on the property on which the action was taken.

The regulatory system provides for relatively minor variations to the general process outlined above in order to accommodate such exceptions as emergencies due to fire or insect infestation; conversions of forest land to other uses (permitted, but subject to regulation); and timber cutting incident to utility right-of-way construction or maintenance, or to Christmas tree, fuelwood, or other minor forest products harvesting, or to removals restricted to dead, dying, or diseased trees. In such situations, the operation is either exempt from the THP review process, or the process is modified appropriately.

Forest Practice Rules

The forest practice rules provided for by the Z'berg-Nejedly act play four important roles: (1) They serve as guides to the RPFs in preparing timber harvest plans, since plans must be so prepared that when they are executed on the ground, the operations are in compliance with the act and the rules. (2) They provide the only criteria that the director may use in denying a plan as not being in conformity with the act. (3) They provide the standards against which on-the-ground operations are checked during law enforcement activities. (4) They provide the basic means for communicating to the public the specific resource protection measures that are expected from private forest owners. Hence, the process of rule development is important.

Proposals for a forest practice rule may originate with landowners, operators, individual RPFs, any one of a number of concerned public agencies, or with the District Technical Advisory Committees (DTAS), the members of the board, or a member of the public. Such proposals are screened by the board's staff, and if they are of sufficient importance, the board determines whether or not it wishes to consider rule making in the proposed problem area. If so, its staff prepares for and announces a public hearing on the proposal.

Staff preparation for the public hearing involes drafting specific language for the proposed rule and preparing a draft public report on the rule. The public report must include (a) the legal authority for adopting such a proposal; (b) a description of the activities affected by the proposal, how these activities would be affected, and the reasons why the proposal is thought necessary; (c) a statement of possible significant adverse environmental effects, if any, that can reasonably be expected from implementing the proposal; (d) a statement of mitigation measures available to minimize significant environmental impacts; (e) a statement of reasonable alternatives to the proposal (usually including the "no proposal" alternative) with pros and cons for each alternative; (f) an estimate of additional costs that would be incurred by operators, landowners, and public agencies if the proposal is adopted; and (g) an estimate of the effect of the proposal on housing costs, if any. Draft language must also meet explicit tests of its consistency, clarity, and adequacy of reference to other sections of law.

The board staff consults widely but informally with representatives of landowners, operators, concerned public agencies, DTACs, and others in preparing these drafts. The draft regulation and public report are published with the announcement for the public hearing at least forty-five days before the hearing. Distribution of the proposed regulation and public report is designed to reach all potentially interested parties. At the public hearing, the board takes formal testimony on the proposal from DTACs, concerned public agencies, representatives of landowners, operators, and other affected groups, and from members of the public. It acts on the proposal in the light of such testimony and, in formal findings adopted to support its action, must respond to each significant environmental point raised at the hearing in opposition to the proposal.

This meticulously defined process, with maximum provisions for both professional and expert input and for expression of public views, means that rules are only adopted after a major effort to provide the board with as much relevant information as possible. But quite clearly, the final decisions rest heavily on the judgments made by individual board members. To adopt a rule, at least five members of the board must vote in support of it. Over the past four years most of the new rules adopted have been approved by more than a bare majority of board members.

Do the complexity of the process and the numerous standards and tests that the rules must meet make the process so cumbersome that rules are rarely changed? In practice, a noncontroversial change in the rules can be accomplished in as little as three or four months from initiation of the proposal. More complex or controversial proposals may require public hearings extending over several months, go through several revisions, and take a year or two to reach fruition.

The outcome of the rule-making process is four sets of forest practice rules, one for each geographic district (or subdistrict). For any one district, the approximately forty typed pages of applicable rules cover such matters as definitions of terms; procedures for estimating such key variables as erosion hazard rating and restocking; applicable silviculture methods; permissible and prohibited logging practices; applicable erosion control methods; provisions for protection of streams and lakes; measures for fire, insect, and disease hazard reduction and protection; responsibilities of landowners, LTOs and RPFs; and procedural requirements of the THP submission and review process.

IMPACTS OF THE INTERVENTION PROGRAM

Let us turn now to an evaluation of this rather diverse program of state intervention in the management of private forest land. The preceding sections indicate clearly that some of these interventions have been aimed at helping private owners cope with various pressures. Other interventions may intensify some of these pressures. Appropriate evaluation should attempt to judge the impact of the entire state policy without focusing exclusively on either the favorable or adverse aspects.

Forest Protection

State intervention to provide, at public expense, basic protection of privately owned forest land from wildfire damage provides a highly significant source of encouragement to investment on such land. Due to its Mediterranean type of climate, California has as severe a fire hazard problem as any area in the world. Critical climatic conditions, which many days each year cause fuel moistures and relative humidities below 10 percent, often accompanied by high winds, are intensified by easy public access to the forests and heavy levels of recreational use. The average number of fire starts on private forest land in California between 1974 and 1978 was more than 300 per year per million acres protected (USDA, Forest Service, 1974-78). The comparable nationwide figure is about 170 per year per million acres. Under these relatively adverse conditions, the average annual burn on private commercial forest land protected is about 0.36 percent. This compares with 0.25 average annual burn on privately owned lands in the United States as a whole.

Broadly speaking, a fire problem about twice as severe as the national average results in an average of only about 40 percent more private commercial forest acres lost to fire each year in California than in the nation as a whole. Losses are kept well below the 0.50 percent level usually considered as a necessary precondition for effective management. These results are achieved largely as a result of state intervention and largely at general taxpayer expense.

In 1977, in the wake of two years of unprecedented drought, northeastern California experienced a major fire disaster in which forests on some 90,000 acres of private land were destroyed. Testimony to basic faith in the efficacy of the fire protection system is provided by the fact that most of the affected landowners immediately began planning for extensive and costly reforestation programs to put the land back in production.

Comparable success cannot be claimed for the state's intervention to provide insect and disease protection. Expenditures on this program have been small (see table 4-2). Losses were high during the 1975-77 drought. To improve insect and disease protection programs, the specialists in these fields need to arrive at a better consensus as to appropriate protective measures. Specialists agree on the need for prompt salvage and for intensive management of young stands, both of which depend mainly on actions in the private sector. Other programs of direct insect and disease control continue to be controversial among specialists and are therefore hardly ripe for general adoption. This, together with the relatively unspectacular nature of the loss, accounts for the fact that state intervention for protection against insects and diseases has been very limited, both in extent and in consequent effects.

Aids to Small Owners

California currently spends about $1.18 million per year (less federal and private reimbursements) on service forestry and in support of state-operated tree nurseries. The service forestry program provides limited advice and demonstration to small landowners, usually stimulating their interest and leading to more extensive service provided by the private consulting sector. As a result the state produces only about 15 percent of all tree seedlings now planted in California.

Before 1971, when there was no significant private nursery output, the state nursery program appears to have made a major contribution to the quite limited reforestation program that was under way at that time. Today the picture is different. Although the state produces only 15 percent of current seedling distribution, much of the private output is either contracted in advance or is for the producer's own account. Thus, the state probably still produces a large proportion of the seedlings annually available on the open market. Since many small private owners (and some industry landowners) do not plan their reforestation programs far enough in advance to contract for seedlings, the state is still performing a significant market service to such owners. The service has economic significance even though current policy is to price state seedlings at the level establised by private operators.

The recently adopted state Forest Improvement Program (with a current appropriation of $4.5 million) has the potential for substantial impact, particularly on long-run timber growth. Since the program is still less than a year old, it is premature to attempt to estimate its impact on individual owners.

Education, Research, and Analysis

State-supported activities provide education for most of the trained foresters needed to maintain the cadre of some 1,800 registered professional foresters in California and (together with U.S. Forest Service activities) contribute most of the research, data collection, and analysis needed to inform forest managers, landowners, and policy makers. In recent years state funds have supported about 50 of the 155 scientist man-years assigned to forestry research by all public agencies (Calif. Board of Forestry, 1979). These functions would have to be carried out at owner expense or not at all if the state did not provide them. If these programs were not supported by the state, private sector efforts in professional education and research would probably be on a much smaller scale. However, most informed people believe that the existing education and research effort needs to be expanded. Thus, these programs in fact seem to be handled more efficiently in the public sector than they would be without intervention. Here it is assumed that landowners benefit from such programs to the extent of the state's cost for them.

Forest Taxation

California intervention in the field of forest taxation has had at least as significant an effect as has fire protection in helping owners cope with pressures on their land. The state's constitutional amendment of 1926 exempted all young or planted timber from taxation until it had reached maturity, or for forty years after removal of the preexisting stand, whichever came later. This appears to have provided a significant shield for

California forest owners against the epidemic of tax reversions that hit the private forest economy between 1925 and 1935. Aggregate reversions between 1911 and 1933 amounted to only 2.1 percent of the area involved in the northern Sierra Nevada (Weeks and coauthors, 1943). In contrast, it has been estimated that 20 percent of the forest area in northern Minnesota and up to 13.5 percent of forested land in timbered counties of Washington and Oregon had reverted to the state. (F. R. Fairchild and Associates, 1935). Such differences in tax reversion may be due in part to smaller proportions of cutover land in the western states, but it seems probable that the favorable tax situation in California also had its effect. Apart from the benefits derived from this intervention during the immediate depression years, California survived the period with its private ownership of forest land very largely intact, in contrast to the fragmentation in ownership that resulted in many other states. This accounts for the fact that California has developed only a relatively small state forest system and has not had to rely on state ownership as a remedy for the instability of private ownership.

Moreover, as immature timber attained a positive market value, the 1926 constitutional amendment precluded inclusion of that value in assessments. This has resulted in some reduction (unmeasured) in the tax burden borne by forest owners as compared with owners of other classes of property.

The adoption of the Forest Taxation Reform Act of 1976 produced several changes. Substitution of a yield tax for the general property tax on mature timber did not (at least initially) affect the tax burden of timber owners compared with that of other classes of taxpayers. So it presumably remains lower (compared with other classes of owners) than would be the case if an unmodified general property tax applied to timber, because the yield tax rate is set so as to raise the same aggregate revenue that had been produced by the general property tax on mature timber in a specified base period. Timber owners, in the aggregate, pay as much total yield tax as they would pay with a general property tax on mature timber. There is some variation in tax incidence among timber owners, however, with the new system favoring those with a lower-than-average ratio of annual cut to mature timber inventory. Of course, for any owner operating on a sustained yield basis, it would make no difference which form of tax applied.

The California forest tax system currently provides three types of benefit to private owners; first, annual taxation of land within the timberland preserve zone (TPZ) rests on assessed values limited to the present net worth of bare land for timber growing, which presumably may be significantly below the general basis for assessment of fair market value; second, a cash flow benefit results from taxation of standing timber at the time of harvest (and thus at a time when income is being generated by the timber), not on an annual basis; and third, owners who do not harvest timber annually benefit through relief from the burden of incurring interest costs on annual taxes on timber from the date of payment to the date of receipt of harvest income.

As of 1978, 5.68 million acres of private land (75 percent) were classified as timberland preserve zone (Calif. Dept. of Forestry, 1979a). The fraction that the timberland assessment on this area actually bears to fair market value varies widely with site quality, extent of recreational development, and other land market factors. In remote areas, the statutory maximum assessment may be close to, or at, fair market value. However, in some areas under development pressure, the assessment on some tracts may be only 10 or 15 percent of what it would be on a market value basis. There has been no survey of the effect of TPZ on assessments. Therefore, more or less well-informed judgment suggests that the valuation of land in TPZ may at present average 80 percent of what it would be on a fair market value basis.

The cash flow advantage conferred by the yield tax feature likewise varies greatly. The large timberland owner who harvests timber each year may receive little or no cash flow benefit from the yield tax as compared with an annual general property tax on all of that owner's timber holdings. But for the small owner who may cut timber only once every ten or twenty years, the cash flow benefit appears to be substantial.

For a representative 500-acre property of average site with a harvest once every ten years, a representative harvest might be 1.75 million board-feet cut from 83 acres. The yield tax obligation at the date of harvest would be $5,740. Using a 10 percent interest rate, the cumulated value of a general property tax on the growing stock of $574 per year over the same ten-year period would have amounted to $7,806. Thus, the yield tax approach in this case reduces the owner's actual cumulated tax burden at the end of the cutting cycle by 27 percent. If the interest saving is annualized and applied to the entire property, it amounts to $0.30 per acre per year. Perhaps of even more importance, the owner has been relieved of the burden of paying timber taxes during the nine years when the forest yielded no income. The capital sum necessary to finance the general property tax on timber in perpetuity would be $5,740.

PRELIMINARY EVALUATION OF THE REGULATORY SYSTEM

Since there appears to be no definitive way of evaluating a system of forest practice regulation, following are five questions that may properly be raised as part of an evaluation of the regulatory system. Tentative answers to those questions are based on available information. These answers are "tentative" because, in practical terms, we have had experience with a number of features of the present system for only a few years. That is enough time to accumulate reasonably definitive data on some aspects of the operation, but not to reach firm conclusions on other important issues, such as long-term watershed impacts, attitudes of owners toward investment, or even success in achieving restocking goals (where the definitive measures need not be made until five years after logging).

Answers to the following questions seem relevant to any evaluation. (The first three are discussed in the sections immediately following, and the last two are addressed later in the chapter.)

    1. How much has regulation done to secure the statutory goals of encouraging prudent and responsible forest management, restoring, enhancing, and maintaining the productivity of timberlands, and achieving maximum sustained production of high-quality timber?

    2. How much has regulation done to secure the statutory goal of giving consideration to the public's need for watershed protection, fisheries and wildlife, and recreational opportunities?

    3. What current costs have been imposed by the regulatory system and how is the incidence of these costs distributed among landowners, timber operators, wood product consumers, and the public treasury?

    4. How has the regulatory system affected the attitude of investors toward committing needed capital to investment in timber growing in California?

    5. How has the regulatory system affected public attitudes which, in the absence of such restrictive regulation, might result in timber-harvesting prohibitions as an alternative to regulation?

Have Statutory Forest Management Goals Been Secured?

Probably few informed observers would deny that current private forest management in California is doing significantly more to maintain and enhance the productivity of forest lands than was the case in 1974. It is equally clear that some of this progress would have been made even without state intervention.

Stumpage prices rose by roughly 275 percent over the five-year period from 1970 to 1974, and by an additional 300 percent over the five-year period from 1975 to 1979 (USDA, Forest Service, 1970-79b). Since these were much bigger increases than the accompanying increases in forest management costs, it is clear that the economic incentives for more capital-intensive forest management were increasing significantly at precisely the same time the regulatory program was put in place. Being of sufficiently unprecedented magnitude, these price rises largely obscured any effects of regulatory factors on the short-run performance of the industry. These same economic forces, however, could also stimulate more extensive liquidation of forest capital. Although the comparative causal importance of regulatory and economic factors may thus be in some doubt, there is little dispute that the response to these two stimuli has been dramatic.

In terms of the policy goal of achieving maximum sustained production, both the immediate and the longer-term situations should be considered. In the immediate term, current annual reported timber products output from private lands in California was 3.7 percent higher during the three-year period after adoption of the Forest Practice Act than in the three years before its passage (Calif. Dept. of Forestry, 1979b). In other comparisons for the same two periods, total domestic timber production declined 0.7 percent, and domestic softwood production remained virtually constant (USDA, Forest Service, 1978). It is of course possible that the better performance in California is only a reflection of better log production reporting.

For the five years after the act was passed, and with data clearly not subject to reporting bias, western softwood lumber production outside California was 2.5 percent lower than during the five years preceding adoption. In California, it was only 2.3 percent lower (USDA, Forest Service, 1980b). Thus, in terms of production response to generally rising price levels, California's performance appears to have been better than that of the nation as a whole and even slightly better than the performance of other areas in the West.

It may be argued that in the absence of forest practice regulation California's short-term production performance might have been even better than it was. Perhaps so. But I can find no hard evidence to support the view that the regulation had a negative effect on short-term supplies.

In this connection it should be pointed out that the production responses just described took place in an increasingly tight stumpage supply situation. Private forest products output in California has been on a generally downward trend since 1953 because of the diminishing availability and accessibility of merchantable stumpage. This trend is likely to continue for another twenty years because of an acute shortage of merchantable age classes, regardless of where regulatory impacts are imposed (Calif. Dept. of Forestry, 1979a). It is quite easy to mistake this inevitable economic contraction for a result of regulation. But the contraction has been long predicted (Vaux, 1955). Regulation did not cause it.

Analysis of effects of regulation on the long-term supply is complex. It seems fairly clear that the restocking requirements of the act are being met on most timber harvest plans by the statutory deadline for filing restocking reports. Only 2 percent of all of the violations of the act noted by inspectors involved silvicultural rules. From 1975 through 1980, some 4,173 stocking reports were received; of these only 118 (2.8 percent) showed inadequate achievement of statutory stocking standards. In any event, deficiencies in stocking thus reported will have to be corrected. As the five-year deadline for filing stocking reports has only expired on land cut prior to 1975, a restocking problem could exist without our knowing it, although it seems unlikely.

On about 25 percent of the plans, restocking requirements are met immediately after logging, usually because selective cutting is being used or because harvesting constitutes thinning. Another 25 percent require artificial regeneration, and about half rely on natural regeneration (Calif. Dept. of Forestry, 1977-79). It appears that about 50,000 acres per year are being harvested in ways that require artificial regeneration in order to meet stocking requirements. Annual data on distribution of planting stock and on areas artificially regenerated confirm that a very rapid and large increase in artificial regeneration has taken place over the period of time that stocking requirements have been in effect.

The distribution of planting stock to private forest owners increased from 5.25 million seedlings in 1970 to 28.25 million in 1979, an increase of 440 percent. The takeoff began in 1972, shortly before the Forest Practice Act was passed. Subsequent to the act, in both 1976 and 1978 there were increases in seedling distribution greater in absolute amount than the total distribution in 1970. Over the entire 1970-78 period, forest industry seedling distributions (mostly for its own use) increased by 15 million, independent nursery distributions increased by 7.50 million, and state distributions remained relatively stable (Calif. Dept. of Forestry, personal communication, Dec. 19, 1980).

Without doubt a part of this increase in seedling production would have come on line without the Forest Practice Act. Some plans for nursery expansion had been announced before the act was adopted. Nevertheless, the restocking requirements of the act are not easily avoidable, and it seems probable that the restocking requirement reinforced the economic incentives for more planting. Also, performance during the years before 1973 had left large areas unstocked, and the existence of this backlog of idle land encouraged the buildup of nursery capacity. The backlog of private land needing regeneration in 1977 amounted to 1.2 million acres, or 15.8 percent of all private commercial forest land (Forest Industries Council, 1980).

The area of land treated to artificial regeneration each year has also been rising rapidly (table 4-4). As has been mentioned, owners have a maximum of five years after harvesting timber in which to achieve the statutorily required minimum level of stocking. To count for purposes of stocking evaluation, seedlings must have been established for at least two years. So operators theoretically have three years to get new trees into the ground. In practice, however, the aggressiveness of competing vegetation is such that prudent managers do their best to plant during the winter immediately following logging.

The total private area planted (or seeded) during 1974 was 16,000 acres. In 1978, 74,100 acres were so treated, which is a four-and-one-half-fold increase in the annual rate. Forest industry planting rates increased fairly steadily over the period. Nonindustrial owners seem to

Table 4-4. Area Reported as Planted or Seeded, by Class of Private Ownership, 1974-79

(thousand acres)

		Class of owner		
	Total	Forestry industry	Other industry	Nonindustry
1974	16.04	11.14	0.53	4.37
1975	25.27	17.81	2.55	4.91
1976	32.14	23.93	2.96	5.25
1977	47.14	37.45	5.13	4.56
1978	59.65	48.63	6.22	4.80
1979	68.87	51.40	8.87	8.60

Source: USDA, Forest Service, Forest Planting, Seeding, and Silvicultural Treatments (Washington, D.C., various years)

have been more cautious, maintaining a low level of about 5,000 acres per year through 1978. In 1979, however, nonindustrial planting rose significantly. We do not yet know whether this reflects the beginning of a new trend. Nor is it known how much is new planting on currently cut areas and thus designed to meet stocking standards, and how much is planting to reclaim the existing backlog of unstocked and understocked areas.

The apparent, striking difference in planting performance between industrial and nonindustrial forest owners may be significant. Although limited, the data support the hypothesis that industrial owners have responded quite steadily to stumpage price increases by planting more and more land each year, while nonindustrial owners did not respond to the price rise, but did begin to react in 1979 when the statutory deadline for meeting stocking standards loomed only a couple of years away.

If this interpretation of the effect of regulation on long-run timber supply is valid, what is the economic implication? Based on the Haynes/Adams model, data for California in 2030 project an annual stumpage consumption of 933 million ft^3 at a price (1978 base) of $2.08/$ft^3$ (Calif. Dept. of Forestry, 1979a). Let us assume that the forest industry contribution to this output will be generated in response to economic factors alone and will not be affected by the imposition of stocking requirements, but that there is some effect on the restocking of nonindustrial land. Some 40 percent of nonindustrial land is now in an understocked condition.

Let us assume that, in the absence of statutory stocking requirements, 40 percent of the currently logged area on nonindustrial lands would not be adequately restocked. Finally, assume that nonindustrial forest lands, which currently account for 64.7 percent of the private land base and 30 percent of the timber removals, will provide 30 percent of the Haynes/Adams projection of the cut in the year 2030.

On these assumptions, we may project that the harvest available on nonindustrial forest lands in 2030 as a result of stocking requirements would be 110 million ft^3 greater than it would be without stocking requirements. If we further assume a price elasticity of demand for stumpage in 2030 of -0.31 (Adams and Haynes, 1980), the assumptions lead to the projections that current stocking requirements would result in an 11.8 percent greater in-state stumpage supply and a 37.7 percent lower stumpage price by the year 2030.

These lower stumpage prices would lead to an estimated 10 percent reduction in consumer prices for wood products (Calif. Dept. of Forestry, 1979a) and a consequent reduction in wood imports. At 0.5 elasticity of supply for imports, the gain of 110 million ft^3 of supply from in-state sources would be partly offset by a loss of 40 million ft^3 in imports. The result would be a 70 million ft^3 larger net supply of wood to consumers at a price of \$.785/ft^3 below that without the enforced reforestation. The annual value of the 7.5 percent larger net wood supply obtained from imposition of the stocking requirement would be \$1.27 billion less than the value of the smaller output obtainable without the stocking requirement.

The present net worth of this saving to consumers has a value discounted at 6 percent of \$68 million per year. Since California is a substantial net importer of wood products and is expected to be much more so by 2030, the stocking requirement feature of the Forest Practice Act appears to have been a significant step toward achieving the legislature's goal of encouraging prudent forest management and achieving maximum sustained production of timber in the long run. Some colleagues have pointed out that if the preceding analysis is correct, it implies a very high cost of intervention to those private owners who are farsighted enough to invest in timber growing in the absence of restocking requirements. Such owners presumably will receive a stumpage price about one-third lower than they would in the absence of regulation. That is certainly a correct interpretation. But, in the face of existing understocked and potentially productive land, those high future prices (and accompanying higher-than-normal profits) are a reflection of severe imperfections in the existing market for forestry capital. It is the express purpose of the stocking requirements of the law to mitigate the effects of those imperfections. The crucial question is not whether regulation reduces the prospect of higher-than-normal future profits for farsighted owners. Rather it should be: Given the existing market structure for forestry capital, will the combination of market incentives and public regulation bring us closer to a long-run equilibrium position for wood products than would market incentives alone? On the basis of evidence presently available, the answer to the latter question clearly appears to be yes.

Have Statutory Goals for Environmental Protection Been Met?

The answer to the question of whether statutory goals for environmental protection have been met is complicated by the impreciseness of the statutory goal itself, which states that the public's need for watershed protection, fisheries and wildlife, and recreational opportunities shall be given "due consideration." In partial clarification, the Forest Practice Act (sec. 4562.7) states further that the board shall adopt rules to control operations that will "threaten to result in unreasonable effects on the beneficial uses of water." In the Porter-Cologne Water Quality Act, the legislature further provided "that the quality of all waters in the State be protected to obtain the highest water quality that is reasonable" (Calif. Water Code, div. 7, chap. 1, sec. 13000), and the act established machinery to provide water quality objectives and to adopt quality control plans to achieve these objectives.

In view of these circumstances, to guide its rule making, the Board of Forestry has a policy under which maintenance of forest productivity and of water quality are given primary and coordinate weight, with somewhat lesser importance attached to fishery and wildlife values, and with a still lower weight for recreational and aesthetic values.

From the enforcement standpoint, the water quality objectives of regulation have proved to be the most difficult to achieve. Violations of the erosion control and stream and lake protection rules typically account for more than half the rule violations reported by inspectors, and the number of such violations has not tended to decline with experience, contrary to what has happened with other rules. However, about 80 percent of the more than 1,000 plans on which logging was completed in 1979 had no violations of any kind during the life of the plan. A considerable portion of the violations on the other 20 percent had to do with forest protection or other requriements and had little environmental impact (Calif. Dept. of Forestry, 1977-79). Given the difficulty of applying rather complex rules to rough, highly variable, and unstable terrain, this is good record of compliance, reflecting a substantial degree of dedication on the part of the concerned landowners, registered professional foresters, licensed timber operators, and logging workers.

In June 1980 the Board of Forestry completed an intensive review on the subject of how well the regulatory system met the requirements of section 208(b)(1)(A) of the Federal Water Pollution Control Act amendments of 1972. The board held extensive public hearings and concluded that the existing regulatory system prevents or reduces the amount of water pollution generated by silvicultural nonpoint sources but falls short of a system of best management practices (as defined in the federal statute) in certain respects. More particularly, it was found that (1) the existing regulatory system controls activities incident to timber harvesting, but does not address potential water pollution arising from nonharvest activities such as regeneration, application of chemicals, noncommerical stand improvement, and long-term road maintenance; and (2) in some instances the rules do not require the most effective practicable means of limiting pollution generated by timber harvesting to a level compatible with water quality goals (Calif. Board of Forestry, 1980c).

During the public hearings, some witnesses testified that with the present rules there is no evidence of water quality problems. Other witnesses asserted that significant problems remain.

A clearer indication of the adequacy of <u>the rules</u> as a means for protecting environmental values is provided by the record of the number of additional measures, recommended by the interdisciplinary team that reviews each timber harvest plan, for mitigating potential adverse environmental impacts. Since the THP should conform to forest practice rules when it is submitted by the RPF, recommendations for mitigating measures indicate the interdisciplinary team's opinion that conformance to the rules is of itself not sufficient protection. On some 3,913 THPs examined in 1977 and 1978, 49 percent resulted in recommendations for mitigating measures; an additional 1 percent were denied as not being in conformity with one or more rules.

One may informally conclude that the present forest practice rules alone merit no better than a "C" grade for achieving the environmental purposes of the act. That is why the board is currently engaged in revising the rules. The system as a whole, however, by providing for additional mitigation based on interdisciplinary team views, is performing significantly better than that, perhaps rating a "B" overall. Present efforts to strengthen the water quality protection provided by regulation will result in still better performance.

Many informed people who have examined current California logging practices have been strongly impressed with the progress made in protecting environmental values. There is little doubt in my own mind that, for the most part, water quality is now being effectively protected; better fish and wildlife habitat is being provided through such measures as limited size of clear-cuts, retention of many snags, and more careful treatment of streamside zones; and there have been some favorable landscape effects in areas of high-intensity recreational use.

As noted earlier, other agencies such as the Water Resources Control Board and the Department of Fish and Game have overlapping authority for some of the environmental protection goals of the Forest Practice Act. Would these other agencies enforce and achieve these goals without the act? If so, why duplicate the efforts? The answer to these questions depends in part on the environmental goal in question. Consider, for example, the case of water quality. The Water Resources Control Board operates by enforcing waste discharge requirements. This concept works best in cases of point sources of pollution. As has been recognized in section 208 of the Federal Water Pollution Control Act amendments of 1972, a preferred approach for nonpoint sources such as silvicultural activities is the concept of best management practice, and such regulations should be developed and enforced by a single agency. Hence, regulation of all environmental aspects through the forestry regulations appears to be more effective than dispersing the process among a number of single-purpose administrative units, both because the regulatory devices available under the forestry law are more appropriate for the purpose and because the goal of "one stop shopping" for permittees is at least attainable.

What Current Costs Have Been Imposed by the Regulatory System?

The cost to the Department of Forestry for administering the Forest Practice Act is currently about $3.1 million per year, as in the state's Budget Act of 1980-81. This represents an average cost per timber harvest plan approved of about $1,150, and an outlay by the state of about $1 per thousand board-feet (bd-ft) of wood output regulated. It does not include the costs of some other state agencies such as the Regional Water Quality Control Boards and the Department of Fish and Game, which, as has been stated, also participate to some degree in the regulatory process.

No data are collected on the costs incurred by private landowners and operators in preparing THPs and carrying out measures required by the rules. Moreover, due to the great variations in California's terrain, vegetation, and soil conditions, highly variable costs among different THPs are to be expected. Even for a THP covering a small area on easy ground with no critical water courses, a minimum cost to the landowner or operator for preparing and filing a THP appears to be about $200 per plan. For larger plans with more difficult problems, plan preparation cost is higher and in extreme cases may exceed $10,000 per plan. If the average cost per plan is $750, the cost for plan preparation is about $.50 per thousand bd-ft harvested. The average cost per plan in the Coast Range (where the most difficult problems are) probably exceeds this amount, with a consequent lower average in the eastern half of the state.

There is a serious data problem involved in evaluating the costs of additional measures that are required by the rules but which would not be used in their absence. The regulatory system should not be charged with the costs of all of the practices specified by the rules. It should be charged only with the cost of those practices the owner would not use if there were no rules. Some industrial foresters have stated that they would do essentially all of the things required by the forest practice rules even if there were no regulation. In such cases, the real cost of the practice requirements of the rules is zero. Since it is impossible to determine what practices operators generally would actually use in the absence of regulation, it seems impossible to measure the marginal addition to cost that enforcement of the rules induces.

As a substitute, I have made some effort to secure estimates in the form of individual evaluations from experienced foresters and operators engaged in logging. Such estimates of the additional costs incurred as a result of practices required by law range from $8 to $35 per thousand bd-ft. As costs are steadily rising and are likely to increase somewhat in the course of implementing water quality standards more fully, a range of $10 to $49 per thousand bd-ft might better reflect the immediate condition. For a representative situation the additional cost probably does not exceed $20 per thousand bd-ft on the average. In the face of current market values of stumpage which range from $90 to $480 per thousand bd-ft (Calif. Board of Equalization, 1979), it seems unlikely that the costs associated with rule requirements will have much effect on the magnitude of operations except in a limited number of marginal cases. Quite clearly, however, the net returns from timber harvesting to some stumpage owners have been significantly reduced.

Another sort of cost arises when the rules preclude cutting timber that otherwise would be removed. Examples are rules requiring that a 300-ft buffer be left around clear-cuts and that the buffer not be cut for three years, and rules that may require leaving 50 percent of the shade-producing crown canopy in stream- and lake-protection zones. In the case of the buffer zones, it would be hard to show that there was any financial loss to owners required to leave timber for three years--between 1975 and 1979, the average stumpage price for all species increased in California at a compound annual rate of over 30 percent. A real cost from such deferrals of cutting could, of course, emerge in the future if stumpage price rises taper off.[1]

There is undoubtedly a real cost to the landowner arising from requirements to leave merchantable timber in the stream protection zones that must be provided for all intermittent and perennial streams. There is no estimate of the magnitude of the actual volume left on this account. Limited data suggest that stream protection zones account for 7 to 10 percent of

---

[1] As of July 1982, the earlier rising trend of stumpage prices had been supplanted by a significant decline.

the forest area. Where selective cutting or commercial thinning are practiced, leave-requirements in stream protection zones probably do not of themselves lead to additional leave-volumes. On the order of 75,000 acres per year appear to be cut by methods that require additional leave-trees in the stream protection zones. Thus, the area involved would approximate 10,000 acres. If the required leave-volume amounts to 5,000 bd-ft per acre of stream protection zones, the aggregate leave-volume would be about 50 million bd-ft per year, the equivalent of a stumpage investment of about $7.5 million per annum.

Presumably, this reserved volume could be removed at the end of the next rotation when younger age classes would be providing the requisite shade. But in the interim, earnings on the reserved growing stock would be significantly below the optimal rate. In addition, it may not be possible to ensure fully stocked stands in the stream protection zone because of its relatively undisturbed condition. Since both the cost and supply impacts of stream protection zones appear to be potentially significant, they need a careful study that should include further evaluation of the effects of stream protection zones on fisheries and water quality.

In summary, it appears that the requirements of the Forest Practice Act involve an additional cash expenditure in logging that may average $55 million ($20 per thousand bd-ft) per year.

This level of cost increase ought to be justified by the value of the resulting benefits in the form of improved water quality, better fish and wildlife habitat, reduced soil erosion, and other environmental values. If such values do result, then the regulations are simply a means of internalizing costs that otherwise are borne largely by nonforest sectors of society. We do not have reliable ways of measuring either the magnitude or value of these kinds of benefits. It can be said, however, that after seven years of operation, the principal political challenge is over whether or not the present system is the most efficient way of doing the job. To date there has been no serious political thrust that challenges either the conclusion that the environmental values are insufficient to merit some form of protection, or that some form of regulation is the most useful approach to such protection.

Most of the $20-per-thousand-bd-ft increase in costs cannot currently be passed on to consumers and must be borne by the stumpage owners in the form of a reduction, due to the forest practice rules, in the immediate harvest value of their timber. In addition, stumpage owners are required to leave perhaps $7.5 million worth of timber standing for an indefinite period for stream protection purposes. Interest on this investment at 10 percent represents an additional cost of $.30 per thousand bd-ft cut.

There is a further, immeasurable, psychological cost. The regulatory system requires a substantial amount of paperwork and fieldwork and places RPFs, LTOs, and forest landowners in a position of being constantly overseen by state officials. This produces very real frustrations and resentments that may be as important in people's lives as the dollar costs involved.

From the public point of view, and with some obvious oversimplification, the economic effect of forest practice regulation in California appears to have functioned during its first six years very much like a windfall profits tax on stumpage with forced reinvestment of the proceeds in environmental protection, especially protection of water quality. Should the unprecedented rise in stumpage values experienced in recent years be checked at some future date, this analogy would of course no longer be appropriate. But I do not know of any informed analysis that concludes that stumpage prices are likely to level off in the future, except for short intervals.

A TENTATIVE OVERALL EVALUATION OF INTERVENTION

This section will provide a tentative evaluation of the effects of state intervention in the forest sector on the various actors.

Costs and Benefits of Intervention to Landowners

How do the cumulative effects of the several forms of state intervention described above influence forest management and forest land use decisions? To answer that question, perhaps the most obvious of several possible approaches is to examine the aggregate effects of all of the programs on the costs of private forest owners. Table 4-5 presents an initial, highly judgmental estimate based on this approach.

The basis for each estimate was discussed earlier in this chapter. In the case of the costs of forest practice regulation, where estimates were made on the basis of costs per thousand bd-ft of production, these

Table 4-5. Estimate of Effects of State Intervention on Costs of Timber Management for Three Types of Private Owners in California (dollars per acre of commerical forest land owned, per year)

Form of intervention	Case I Small owner, land in TPZ Credit	Case I Debit	Case II Small owner, land not in TPZ Credit	Case II Debit	Case III Industrial owner, land in TPZ Credit	Case III Debit
Forest protection	3.20	--	3.20	--	3.20	--
Education, research	0.50	--	--	--	0.50	--
Service	0.20	--	--	--	--	--
Taxation						
Reduced land tax	0.30	--	--	--	0.10	--
Reduction in carrying charges	0.30	--	0.30	--	--	--
Forest Practice Regulation						
THP preparation	--	0.40	--	0.40	--	0.20
THP execution	--	7.40	--	7.40	--	6.00
Stream protection reserve[a]	--	0.50	--	0.50	--	0.50
Total per acre	4.50	8.30	3.50	8.30	3.80	6.70
Net balance per acre[b]	--	3.80	--	4.80	--	2.90

Notes: Dash = not applicable. TPZ = timberland preserve zone. THP = timber harvest plan.

Methods of deriving estimates are described in the text. Figures have been rounded to the nearest $0.10 to avoid unjustifiable impression of precision.

[a] Cost varies from $0 under selective management to $3.15 per acre under clear-cutting. Entries reflect a weighted average of cutting methods.

[b] Credit for federal subventions, not included in table, but which provide additional support from some listed programs, amounts to $0.13 per acre.

were converted to a per acre basis by multiplying them by the average annual production per acre of private commercial forest (370 bd-ft per acre per year).

Three types of situation are presented in table 4-5. Case I reflects the situation for a small nonindustrial owner who has 500 acres of average site land included in timberland preserve zone (TPZ), and who is presumably engaging in forest management. It is assumed that the owner makes one sale of 1.75 million bd-ft every ten years. Case II represents an owner whose land is not in TPZ and who does not engage in forest management, but who otherwise resembles case I. Case III represents a forest industry that owns 150,000 acres in TPZ. Debits on account of regulatory costs are higher for small owners than for industry because of assumed economies of scale in the larger enterprise. Case II owners do not benefit from service forestry, research, and education, so no credits from this source are shown.

Following are comments on the net balances shown in table 4-5. Larger industrial owners appear to be penalized less than small owners, as a result of economies of scale that permit them to absorb regulation requirements at less cost. However, economies of scale would also presumably affect forest protection cost, with the small owner being relieved of a heavier cost burden than the large one. Since there is little basis for judging economies of scale in fire protection, they are not reflected in the table, but they might well be sufficient to reverse the comparative position of large and small owners.

Small owners in TPZ who manage their land for timber production are less adversely affected by intervention than those owners not in TPZ who do not manage their lands. This is, of course, consistent with the state's policy objective of encouraging prudent forest management.

Table 4-5 shows that the case I owner experiences an annual net debit, on account of state intervention, in the amount of $3.80 per acre owned. However, the table allows no credit for the newly instituted state Forest Improvement Program. As already noted, this program provides $4.5 million per year for cost sharing; about 3.6 million acres of land are eligible for the program. This works out to $1.30 per eligible acre. If successful, this program will in the aggregate redress about 35 percent of the debit balance incurred by case I owners. It is of course true that the owners who will secure forest improvement contracts are not necessarily the same ones who are incurring the costs of regulation. In this sense intervention tends to favor the owner whose lands are in poor shape and who has opportunities for forest improvement, over the owner whose land is already fully stocked.

But that is not the whole story. Several real economic effects of state intervention cannot be shown in balance sheet form. These include:

1. Financial advantages to some owners from the favorable effect of the yield tax on their cash flow position, compared with the effect that would result from an unmodified general property tax on timber.

2. Financial advantages to some owners from the existence of a state-supported nursery system that bears the risks of producing seedlings for an unpredictable market and must absorb losses when such markets do not emerge.

3. Financial advantages from state income and inheritance tax modifications, for which evaluative data are not available.

The aggregate of these advantages may offset part, if not all, of the negative balance as far as some owners are concerned.

With these points in mind, let us relate the reduction in annual net returns per acre suggested by table 4-5 to current land values. By comparing land values used for taxation purposes (Calif. Board of Equalization, 1979) with the net debits of table 4-5 capitalized at 10 percent, one may judge whether or not land has been driven out of forest production by state intervention, with its presumed adverse effects on net returns. The burden is above the value of Site Class V Land for all owner classes shown in the table, but it is below the value of all the higher site classes for all except low Site Class IV lands and case II owners. The latter are not in TPZ, and in any case it is probably not realistic to consider their land as part of the long-term timber supply base. Thus, the principal effect we are concerned with is the potential of the intervention program for driving Site Class V lands out of production. These lands are generally considered to be below the economic margin for timber production. Moreover, at full production they would contribute less than 2.6 percent of the potential productivity of all forest lands in California.

Recognizing the tentative and rough nature of some of the cost estimates, it is concluded that for the average forest landowner the balance of direct financial impacts of state intervention is not of itself likely to have a major effect on land management and land use decisions.

There are, of course, legitimate equity concerns raised by the net debits suggested in table 4-5. These concerns result from the reduction in the value of forest owners' land implied by these debits. They range from $29 per acre to almost $50 per acre, if one ignores the financial advantages, noted above, that are not reflected in the table. As economics lacks generally acceptable criteria for judging equity where intergenerational income distribution problems are at issue (Page, 1977), no judgment is made here as to the propriety of the present example.

Forced Reinvestment

As indicated earlier, the evidence seems fairly clear that the restocking requirements of the Forest Practice Act force some owners who would not do so otherwise to reinvest in timber growing. Based on data and assumptions outlined earlier, an estimated investment of perhaps $2 million to $3 million per year, forced by the restocking requirements of the act, would not have taken place under pure market incentives. We do not know what alternative lines of consumption or investment have been curtailed by this requirement.

Most of this forced investment takes place on high Site Class IV land or better. Studies have shown that such lands are capable of producing a real return of 6 to 10 percent per annum or better on restocking investment (Vaux, 1972; Forest Industries Council, 1980). The legislative policy requiring this reinvestment was adopted explicitly to correct perceived imperfections in the capital market as it affects forestry. Therefore, we cannot conclude that this forced reinvestment represents an uneconomic diversion of capital, at least on any significant scale.

It does, however, have significant income distribution effects. Broadly speaking, it redistributes income from the present generations of landowners to generations of consumers that will enter the market in 2030 and beyond. It is a rather pure example of what Ciriacy-Wantrup defines as _conservation_ (Ciriacy-Wantrup, 1973). As was indicated earlier, the present net worth of the income thus distributed to future consumers is on the order of $70 million per annum. In the absence of an accepted intergenerational income distribution criterion, one cannot say that there is a net economic gain from this income redistribution effect. But considering the general outlook for future supplies of natural resources, the estimated advantage to future generations appears to significantly outweigh the present costs.

Psychological Costs

Even if, as has been suggested, the effects of state intervention on the financial profitability of timber growing are limited and the resultant intertemporal income redistribution is justified, a similar conclusion for real psychological effects of the program does not necessarily hold. The markets in which timber-growing investors operate are, I submit, far from perfect markets. Some of the imperfections and their potential impacts on management and land use can be illustrated by reference to case III of table 4-5.

The debit item of $6.70 for forest practice regulation represents landowners' out-of-pocket cash expenses resulting from intervention. On the credit side, the taxation items represent a reduction in what taxes might be otherwise and the protection item represents provision of a state service that landowners have been accustomed to enjoy for more than fifty years. It seems entirely likely that cash outlays to meet regulatory requirements have a stronger effect on owners in a psychological sense than do hypothetical tax and protection savings. Moreover, the protection and tax interventions have been on the books in some form for half a century, and when such considerations are raised, the forest owner is as entitled as the rest of us to say, "I know, but what have you done for me lately?"

A final psychological cost--and it is a real cost in the sense that it may influence owners' behavior--is the frustration that attends even the best-conceived and best-administered regulatory process. None of us wants government intervening in our day-to-day operations, and such a program thus encounters resistance (cost) above and beyond the monetary ones.

Some of these psychological costs are likely to be reduced as the participants in the system become more accustomed to new and previously untested relationships. The Board of Forestry and the Department of Forestry can reduce other costs of this kind by the tone and good judgment given to administration of the law. Such cost reductions will take time to achieve, but they are potentially attainble. A psychological problem of a different order is evidenced by the frequently heard landowner's question: "How do I know I will be allowed to harvest my timber after I have grown it?" State intervention is only one of the causes of this problem, and to date local government and federal land use decisions have probably been more significant causes of this particular uncertainty. But state regulation adds fuel to the fire. This issue will be discussed in the broader context of the next section.

How Has the Regulatory System Affected the
Climate for Timber-Growing Investments?

Let us now turn to the fourth of five questions, posed earlier in this chapter, to be considered in evaluating the regulatory system. The question is how the regulatory system has affected the attitude of investors toward committing needed capital to investment in timber growing in California.

It has often been suggested in recent years that the $55 million per year negative impact on the financial position of stumpage owners is having (or may soon have) a devastating effect on the climate for forestry investments. More specifically, many forest landowners and others have testified that unless the regulatory burden can be reduced, prospective investors will turn away from timber-growing investments that otherwise might have been made in California (Calif. Board of Forestry, 1980a; Forest Industries Council, 1980).

This is a very serious consideration. The California Forest Productivity Report, 1980 (Forest Industries Council, 1980) estimates that $366.7 million of new investment would be required to realize those opportunities for increased growth on private land in the state that would yield a return on the capital invested of 10 percent after taxes. For reasons similar to those analyzed in discussing the consumer benefits from restocking requirements, realization of these potential private investment opportunities is important. They would result in additions to future supply three or four times the size of those previously attributed to the restocking requirements. If it is true that the regulatory system is shaking the confidence of private investors to the extent that most of this needed new investment will be inhibited, it would raise most serious questions as to the wisdom of the current regulatory policy.

The evidence in this matter consists of both objective data and essentially anecdotal information. The former is the record of new investment in reforestation which, as already noted, has increased strikingly. The latter evidence, which is fragmentary and conflicting, takes the form of the testimony of potential investors.

The operator on one medium-sized operation evaluated his experience with the Forest Practice Act by saying that his business had never had it so good. In contrast, executives for a major international wood products producer said they would never invest in timber growing in California so long as the present regulatory controls remained in effect. One major company has sold its California operations since the Forest Practice Act went into effect, ostensibly because it judged that its money could be better invested in other areas. However, as soon as the property was put up for sale, it was purchased by another out-of-state concern at a record price. Another large land-owning company lost a substantial portion of its forest land holdings to federal condemnation for the Redwood National Park. The company had particularly difficult experience with the Forest Practice Act resulting from pressures generated by the fact that some of its timber was very close to the park. Subsequent to federal condemnation of its land, that company purchased more than 20,000 acres of forest land in the redwood region to help enlarge its depleted land base. Other companies are currently engaging in significant investments in new tree nurseries, new and modernized utilization plants, and in other activities that appear to reflect a continued commitment to industrial forestry in California.

Increased costs of the magnitude I have suggested as being induced by the Forest Practice Act will indeed somewhat dull the edge of investment enthusiasm, and there is some basis for investors' perceptions of possible future restrictions on their ability to cut timber. But there is little evidence that, in the case of the forest products industry as a group, the regulatory system has checked investment to a degree that will preclude further rapid intensification of industrial forest management in California.

For the owner of the small nonindustrial forest, the situation may be somewhat different. It is clear that the burden of complying with forest practice regulations falls more heavily on nonindustrial than on industrial owners. But, prior to the act there had been little investment in timber growing by such owners. It is too early to say whether regulation has checked a surge of new small-owner investment in forestry. But it may be noted that other regions seem to face problems of low investment on small ownerships even in the absence of California-style regulation.

The fact remains that investment psychology is a very important and notoriously sensitive phenomenon. We still have much to do to persuade landowners that what the Forest Practice Act requires of them is (in the light of the state's investment in fire protection, tax equitability, and research and education) both reasonable and constructive.

Efficiency of Reallocation in the Direction of Environmental Values

The reallocation of resources accompanying California's intervention policies clearly appears to have had favorable effects on those environmental values recognized in the regulatory program. Probably the most important of these benefits in terms of social value are the improvements in water quality and the reduction in soil erosion losses as compared with pre-regulation conditions.

The major question here is whether the existing regulatory system is the most efficient way of reaching the stipulated water quality and soil protection goals. There are two types of obstacle to answering this efficiency question. First, the existing data are very sketchy with regard to the actual physical effect of many required forest practices on water quality and on soil movement. This is so because of the widely varying performance under different field conditions. The cost of securing such data appears to be high, and in some instances by the time the data were available, significant losses might well have occurred.

The second obstacle derives from the limited nature of the efficiency concept. It can only be applied where a single objective function is specified or where the appropriate trade-off ratios between multiple objectives are given. The statutory objectives of the regulation program provide for "due consideration" of the several values, not for explicit weights that would permit application of the efficiency criterion.

A different sort of problem arises from the fact that standards of water quality and standards for reforestation, silvicultural performance, and other forest management activities have been derived independently of one another (Calif. Dept. of Forestry, 1979a). Thus, there is no assurance that the legally required standards are consistent with one another. Where they are not mutually consistent, evaluation faces a kind of Catch-22 dilemma.

Under these circumstances, and in accordance with the concept of best management practices defined in section 208 of the federal Water Pollution Control Act amendments, California's approach has been to apply practices which, on the basis of extensive public testimony and in the view of the Board of Forestry, appear to achieve the goals best. These best management practices are subject to continued monitoring and, if necessary, to revision in the light of operating experience. This is not a particularly elegant or even comfortable solution. But so far it seems to have elicited a generally cooperative and constructive effort by all concerned to develop solutions that will stand the tests of experience.

The net effect appears to have been a significant improvement of logging performance insofar as it affects both water quality and soil stability, with no more than marginal adverse effects on current timber output.

Income distribution effects like those noted in the case of forest regeneration appear to result from the imposition of water quality and soil stabilization controls. But numerical evaluation of these effects would encounter the same physical data base problems noted in the discussion of the efficiency of these controls.

How Has the Regulatory System Affected Public Attitudes?

The last of the five questions posed earlier with regard to evaluating the regulatory system involves the effect of the system on public attitudes. An implicit assumption of the cost estimates made above is that it would be possible to operate very much as usual even if there were no forest practice regulation. I believe that is to some degree an untenable assumption. In California, the realistic alternative to operations under the Forest Practice Act is not unregulated operations, but rather a situation in which, over time, timber harvesting may be entirely proscribed over a larger and larger area.

The present regulations are, in fact, a compromise between a position of less restrictive timber cutting and one of still more restrictive and more costly restraints. The tensions over this issue are most clearly evident at the growing interface between commercial forest operations and expanding residential settlement.

In two heavily urbanized counties in the San Franciso Bay region, timber harvesting has been virtually banned for a number of years. Steps to make the ban formal and permanent are currently being pushed vigorously. Several other urbanizing counties have ordinances that go substantially farther than the Forest Practice Act in placing constraints on timber harvest. The primary unresolved question in these expanding interfaces is whether local controls more adverse to timber growing will be adopted, or whether the state regulatory system can be managed in a way that will defuse urban concerns over timber harvesting, but still permit timber growing to continue as an economic use of the land.

Political pressures to curb timber harvesting (and hence timber growing) in California were acute in 1973 when the Forest Practice Act was adopted. Pressure to impose additional constraints still exists. But it appears that the existing policy as conducted by the Board of Forestry over the past few years has created a more stable outlook. I expect that continuation of these policies will reinforce public opinion that sees timber harvesting as an economic activity essential to the state's welfare and as an activity that can be, and is, conducted with reasonable concern for and protection of essential environmental quality elements that Californians clearly wish to defend.

If this perspective is adopted instead of the more conventional one, that is, the alternative of no regulation, which I believe is totally unrealistic politically, then the economic conclusions suggested above need to be substantially revised. What were treated there as net costs to forest landowners and the state would have to be considered as investments, designed to ensure a climate that would permit long-run perpetuation of the forest industries in California. The benefits of achieving such a climate lie in fuller supply of forest-based consumer goods and services, broader employment opportunities, the existence of an economic basis for a significant form of open-space land use, and a much fuller and varied utilization of the basic productivity of the forest resource than would otherwise be possible. Looked at from this perspective, I believe the potential net economic gain to all the sectors of the California economy is clearly far in excess of the costs.

REFERENCES

Adams, D. M., and R. W. Haynes. 1980. *The 1980 Softwood Timber Assessment Market Model.* Forest Science Monograph No. 22. *Forest Science* (Suppl.), vol. 26, no. 3.

Arvola, T. F. 1976. *Regulation of Logging in California: 1945-1975* (Sacramento, California Division of Forestry).

Bayside Timber Co. v. Board of Supervisors of San Mateo County. 1971. App., 97 *Cal. Rptr.* 431. Sacramento.

California Board of Equalization. 1979. *Values for Timberland Zoned as TPZ.* Rule No. 1025, C.A.C. adopted Dec. 11, 1979. Sacramento.

California Board of Forestry. 1979. *Wildland Research Program Needs and Policies for California.* Report of Standing Advisory Committee on Research Programs. Internal report (Sacramento).

──────. 1980a. *California Private Forests: Policies for Renewed Productivity.* Report of California Forest Improvement Committee (Sacramento).

──────. 1980b. *The Market for Tree Seedlings in California.* Internal report (Sacramento).

──────. 1980c. *Report of the Board of Forestry to the Water Resources Control Board.* Adopted June 11, 1980 (Sacramento).

California Department of Forestry. 1977-79. *Annual Report: Forest Practices.* Annually (Sacramento).

──────. 1979a. *California's Forest Resources: Preliminary Assessment* (Sacramento).

──────. 1979b. *Production of California Timber Operations in 1977.* State Forest Note No. 74 (Sacramento).

California Legislature. 1966. *Transcript of Proceedings.* Assembly Subcommittee on Forest Practices and Watershed Management. Hearing on Forest Practices and Watershed Management, Aug. 17, 1966. Sacramento.

──────. 1972. *Transcript of Proceedings.* Assembly Committee on Natural Resources and Conservation. Hearings on Comprehensive Watershed Management and Forest Practices. Eureka, Jan. 14, 1972, and Sugar Pine Point, Sept. 5-6, 1972. Sacramento.

Ciriacy-Wantrup, S. V. 1973. *Resource Conservation: Economics and Policies* (Berkeley, University of California Press).

Clar, C. R. 1959. *California Government and Forestry* (Sacramento, California Division of Forestry).

Forest Industries Council. 1980. *California Forest Productivity Report, 1980* (Washington, D.C.).

F. R. Fairchild and Associates. 1935. *Forest Taxation in the United States*. USDA Misc. Pub. No. 218 (Washington, D.C., U.S. Department of Agriculture).

Natural Resources Defense Council v. *Arcata National Corp*. 1975. 59 C.A. 3rd 959, 131 *Cal. Rptr*. 172.

Nelson, DeWitt. 1952. "Progressive Forest Management in California," *Journal of Forestry* vol. 50 (April), pp. 259-261.

Page, Talbot. 1977. *Conservation and Economic Efficiency: An Approach to Materials Policy* (Baltimore, Johns Hopkins University Press for Resources for the Future).

University of California, Davis. Institute of Ecology. 1972. *Public Policy for California Forest Lands*. A report to the Assembly Committee on Natural Resources and Conservation (Davis).

USDA, Forest Service. 1954. *Forest Statistics for California*. Forest Survey Release No. 25 (Berkeley, California Forest and Range Experiment Station).

———. 1970-79a. *Forest Planting, Seeding, and Silvicultural Treatments in the United States*. Annually (Washington, D.C.).

———. 1970-79b. *Stumpage Prices for Sawtimber Sold from National Forests, by Selected Species and Regions*. Annually (Washington, D.C.).

———. 1974-78. *Wildfire Statistics*. Annually (Washington, D.C.).

———. 1978. *The Demand and Price Situation for Forest Products: 1976-77*. USDA Misc. Pub. 1357 (Washington, D.C.).

———. 1980a. *Production Prices, Employment and Trade in Northwest Forest Industries. Third Quarter, 1979* (Portland, Pacific Northwest Forest and Range Experiment Station).

———. 1980b. *Production, Prices, Employment and Trade in Northwest Forest Industries. First Quarter, 1980* (Portland, Pacific Northwest Forest and Range Experiment Station).

Vaux, Henry J. 1955. *Timber in Humboldt County*. Bulletin No. 748 (Berkeley, California Agricultural Experiment Station).

———. 1972. "How Much Land Do We Need for Timber Growing?" *Journal of Forestry* vol. 71, no. 7, pp. 399-403.

———. 1980. "Urban Forestry: Bridge to the Profession's Future," *Journal of Forestry* vol. 78, no. 5, pp. 260-262.

Weeks, David, A. E. Wieslander, H. R. Josephson, and C. L. Hill. 1943. *Land Utilization in the Northern Sierra Nevada*. Special publication of the Giovinni Foundation of Agricultural Economics (Berkeley, University of California).

William J. Moshofsky

COMMENTS

I welcome the opportunity to comment on the subject of state regulation of forest practices in general, and in particular on California's regulatory system. While forest practice regulation is mostly limited to states on the West Coast at present, other states are considering more regulation, so discussion of the issues involved is very timely and very much needed.

Furthermore, the entire subject of regulation is receiving much attention in Washington, D.C., right now. There is a growing view that overregulation, including environmental overregulation, is a major cause of the current economic ills of the nation. Interior Secretary James Watt has been spotlighting the environmental overkill on federal lands that has greatly restricted resource exploration and development.

The California forest regulatory system is, I believe, the most pervasive, restrictive, and costly system in the country--and I believe unnecessarily so. Clearly these views are in strong contrast with those of the author of chapter 4. But then, Henry Vaux heads the Board of Forestry in California, and the company I work for is one of the rulees. As Vaux indicates, the Board of Forestry and the Department of Forestry have enormous power and discretion over our forests and forest operations. We are literally at his mercy, since the court system and the political system, heavily dominated by urban values, offer the forestry sector little protection. I should also add that there are serious controversies pending in California over substantive as well as procedural aspects of the current system.

I would like to present these comments under a flag of truce, and hope Henry Vaux will not take offense at my taking issue with his comments, for

I do believe there is value in candid, straightforward comment on both sides, and that Californians as well as people in other states will benefit from thoughtful dialogue on the California system.

The enabling legislation for the California forest regulatory system delegates a great deal of discretion to the state forestry board and department--not only as to the content of the rules promulgated, but as to the priority to be assigned to concerns such as timber growing and harvesting, water, soil, wildlife, recreation, aesthetics, and so forth, and as to how rules are enforced.

One of the most troublesome elements in the California system is the requirement that there be prior approval of timber harvest plans by the Department of Forestry. With any government permit system there are risks of delays, of excessive bureaucracy, and worst of all, of the imposition of unwarranted requirements as a "condition" for receiving a permit. But probably a more basic problem with the California system (in fact with all land use regulatory systems) is that regulations can impose costly restrictions on land use without requiring the regulatory agency to reimburse the landowner for the cost of compliance.

The rationale for such uncompensated use restrictions is that they are being imposed under the so-called police power of government, that is, under the power to impose regulations to protect public health, safety, and welfare.

No one has any quarrel with exercise of the police power properly applied--for example, to control highway activity, assure plant safety, and prevent pollution. In those situations, no one would urge the regulator to pay damages to the regulated. However, when the real purpose and effect of regulation is not primarily to prevent harm but to transfer benefits from one property owner to another, or from one property owner to the public or to certain members of the public, a strong case can be made for requiring compensation to be paid to the property owner.

Unfortunately, the courts have been permitting serious abuses of the police power. Even though courts do, without hesitation, require just compensation when there is a "physical taking," they waffle when the taking is achieved by restricting use by regulation. No compensation is required if they find that the regulation serves a "public interest," that it is "reasonable," and that the property owner is left with some use. Indeed, the California Supreme Court says no compensation is payable no matter how restrictive the use, and that the only remedy of the landowner is to sue to have the regulation declared invalid. The case is on appeal to the U.S. Supreme Court.

Against this background, let us consider the California forest regulatory system which is, in my opinion, a prime example of serious abuse of the police power--abuse not only of rights of forest landowners but abuse of the public interest in wise use of resources to provide products, job and tax revenues for schools, communities, and so forth.

EXAMINING CONCERNS ADDRESSED BY REGULATIONS

According to Vaux, the purpose of the California forest regulatory system is to assure "a socially adequate supply of timber" together with "a socially acceptable degree of environmental protection." Rather than dealing with that general statement--which admittedly sounds good--let us break out the concerns that the regulations are intended to address, and examine each one critically. These concerns are primarily reforestation, water, soil, wildlife, recreation, aesthetics, endangered species, and coastal concerns.

Reforestation

Reforestation concerns are primarily economic; they have little to do with environment. Some may say that if government's "economic" decisions on reforestation expenditures are different from owners' decisions, the government should pay the extra cost. Personally, I have real problems with political decisions, as opposed to marketplace decisions, on economic matters. Vaux attempts to rationalize intervention as an externality of the marketplace. I disagree, but cannot spell out my reasons in the space of these brief comments.

However, in the West, forest landowners have tended to go along with mandatory reforestation measures. At least in this regulatory situation, the benefits flow primarily to the landowner--presumably the value of the land is enhanced. Also, regulators have been requiring little different from what owners should be doing anyway pursuant to market forces.

Water

Water concerns are far different from those of reforestation. They involve various aspects, including: (1) volume and flow of water in streams and rivers; (2) quality of the water as impacted by forest operations; either fish or drinking water or both are affected; (3) siltation caused by soil runoff, which may affect fish habitat; and (4) broad concerns such as those of watershed and flooding. Each of these concerns calls for somewhat different approaches. Clearly there is some basis for some regulation relating to water. But the extent is debatable. The water in streams and rivers is considered publicly owned; thus, regulations to protect such water are simply protecting public rights. But the public rights are not absolute. They must coexist with the rights of adjoining landowners to use and enjoy the land. Neither is exclusive.

Furthermore, with respect to the banks of rivers or streams and the land along them, the case for government regulation becomes more tenuous. There may be a public right, based on local law, to undisturbed banks and beds, but this right certainly does not extend to the adjoining land. The right to regulate the adjoining land must be based on some very clearly proved relationship of activities on the land to a legitimate public right in the water.

In this connection, there has been much attention to water temperature as affected by shade. I submit that the public really has no right to a particular water temperature or to the vegetation along the banks. This concept simply evolved and became accepted, partly because our industry went along, never challenging the concept. At the outset it was not too important, but now regulations are extending the buffer areas 200 feet deep, and under wild river protection concepts, a quarter of a mile. At some point, the right to restrict by regulation without paying compensation to landowners must end, or the right to private property will become meaningless.

Obviously, in some situations regulations may be warranted to prevent flooding and excessive siltation, but again, such concerns should not be the basis for excessive regulation.

What is disturbing about water concerns is that Vaux says they have equal standing with timber growing and harvesting. While some aspects of water may deserve equal standing in some circumstances, I seriously question such an approach for others, for example, streamside protection, water quality, watershed, flood prevention, and siltation. The potential for abuse in these situations is enormous, particularly as there are many environmental groups pushing special interests by seeking to restrict harvest (for whatever reason) under the "public interest" label.

I cannot help but think that concerns over water in many areas are greatly overstated. I have seen much of the coastal area after brutal logging done years ago. The streams and rivers have recovered so well that people want to preserve them just as they are. Forest practices, even without regulation, are much less brutal. Perhaps the real public interest can be protected with much less pervasive regulations.

Soil

Concern over soil is closely related to water concerns. Prevention of soil loss by regulation is based on the old legal concept of "waste," which was unacceptable destruction of one's property. This is a common concern of property owners and of the public. Any regulation must be reasonable and even-handed--why, then, should there be pervasive regulation of forest land where soil disturbance is very infrequent, and so little regulation of farmland where there is such frequent soil disturbance?

Wildlife

The wildlife concern is entirely different. Here the basis for regulation is very limited. The public does own some wildlife (such as deer, elk, and other species that roam), but legally, forest landowners have no more obligation to provide food or habitat for such wildlife than farmers do. Fortunately, such wildlife is compatible with forest management, and forest landowners generally accommodate wildlife needs voluntarily. But there is growing anxiety that wildlife concerns will be used to justify more forest regulations without compensation to landowners.

I am not sure whether wildlife concerns are public interest or special interest concerns, but it seems that at some point either the public or the special interests should bear the cost of restrictions imposed on forest landowners for wildlife purposes. The danger of "free" regulation is the potential for abuse. If 500 acres of habitat are free, why not 5,000 or 10,000 acres?

Aesthetics and Recreation

Recreation and aesthetic concerns probably have the least basis for forest regulation, certainly for uncompensated regulation. I believe everyone would agree the public has no right to use the private property of others for recreation, and no <u>right</u> to require forest landowners to farm their forests in a particular way to achieve aesthetic objectives.

Aesthetics is a nebulous, subjective concept in any case, and many believe that in the long run, a managed forest provides the best aesthetic value--certainly better than those provided by nature's crude tools of fire, insects, and old age.

Even if there is a basis for regulation, it must expressly include compensation to the landowner. If the public likes it, they should buy it, so that not all the burdens fall on the forest landowner, and not all the benefits with no burdens go to the public or some members of the public. Without the compensation restraint, the potential for abuse is immense.

I agree that we need much public education to increase people's toleration of temporary unsightly harvesting scenes.

Other Concerns

Although it is not possible to cover endangered species, wild river, and coastal zone concerns here, the same principle applies: absent some clear need for protection of public rights, or prevention of harm, there is little basis for regulation. If there is regulation, the owner should be compensated for the cost of compliance. The public, or the people who benefit from the regulation, should bear the cost. This approach not only protects property rights, but is basically fair and protects the public interest in wise use of resources for products, jobs, tax revenues, and so forth.

JUSTIFICATIONS FOR FOREST REGULATION

Before closing, I want to respond to three justifications for forest regulation suggested by Vaux. The first is that the state is somehow entitled to impose such regulations because it has provided fire and insect prevention and protection programs, some aid to small landowners, and some variation in taxes. I disagree. Each program should stand on its own and should not be justified ex post facto. Furthermore, those state programs are applied unevenly, benefiting some owners and not others.

The second justification is that the Environmental Protection Agency (EPA) requires regulation. That is not true. EPA has limited authority. It has approved <u>voluntary</u> forest practice systems in other states.

The third justification is that regulations are needed to appease the California public, which is more than 90 percent urban, to head off outright prohibitions against forest harvesting. This is frightening, and furthermore it is a sad commentary on the understanding and rationality of the people and on our political and legal system. If the legislature, state agencies, and courts will not protect the economic base of the state as well as the fundamental rights of the minority against the whims of the majority, our country is in serious trouble. If public opinion is so powerful, and our politicians and judges so impotent, it is time for a concerted effort by the industry, labor, consumers, communities, forestry schools, and others to educate the public about the realities of forest management and the stake the public has in wise use of those resources.

On the "stake" question, I want to stress that the costs of the California forest regulatory system are high and going higher. Our company administrative costs alone are about $14.00 per thousand bd-ft (compared with $1.00 to $1.50 per thousand bd-ft otherwise) to comply with the current system. This does not include extra logging costs, extra road costs, and the value of trees required to be left. Our industry believes that many of these rising costs are needlessly imposed.

The forest industry is not opposed to necessary regulations. We simply ask that they be reasonable, and that they be enforced in an efficient and effective way--and that the real benefits justify the increasing costs.

Chapter 5

REGULATION OF PRIVATE FOREST LANDS:
THE TAKING ISSUE

M. Bruce Johnson

The federal Constitution and a number of state constitutions contain the provision that private property may not be taken for public use without just compensation. Taken at face value, this provision is the citizen's constitutional guarantee that his or her property rights will be protected from arbitrary expropriation by government--or at least that the person will be compensated if the property is taken. Most of us would agree, I believe, that the principle of the just compensation provision is sound, for each of us can imagine a government captured by special interests (other than our own) or by well-intentioned visionaries whose master plan is based on an insatiable appetite for resources far beyond the public budget.[1] Such a view may sound cynical, but I submit that it is a realistic interpretation. Indeed, the very existence of constitutions implies an intrinsic distrust of governmental powers.

The compensation (or "taking") clause sounds straightforward enough until one begins to search for operational definitions of terms like "private property," "public use," "just compensation," and "taken." The essential complication arises when the practical alternative to taking is

---

[1] "Few would disagree that if the government rezoned a tract of, say, 100 forested acres zoned for high-rise apartments to public park and allowed no other use of this land, it should compensate the owner for the full value of the property. Although title to the land may remain with the owner, the public has, for all practical purposes, assumed ownership. Zoning that effectively classifies property for a public park, school site, highway, or public building will be struck down [footnoted case omitted]..." (Siegan, 1977, p. 18).

considered: acting under the police powers thought to be inherent in the nature of the state, the government may regulate the use of private property without the duty to compensate the property owner. So the issue becomes a question of when a government should "regulate" and when it should "take."

Because taking requires compensation and regulation does not, the question is far from trivial. Indeed, the issue is fundamental to the kind of society we have had and expect to have. Social, political, and philosophical as well as economic implications are involved in the fundamental question of who will be entitled to decide how resources will be used. Will individuals and groups be free to trade claims to the use of resources voluntarily, or will the government monopolize claims to resource use? Will the right to decide resource use rest with individual property owners, subject to a case-by-case showing by the government that the public interest is better served by government intervention? Or will the rights belong to the government, with the burden of proof on individual property owners to show why and how on a case-by-case basis they should be allowed to use their property? Our history reveals a tug-of-war between the two views; the first view predominated in the nineteenth century, and the latter predominates today.

The origin of the recent renewal of interest in the taking issue is benign enough. On the one hand, we have witnessed a heightened awareness and concern for the environmental consequences of economic activities. On the other hand, many property owners are increasingly concerned that their entitlements to use their properties will not be respected and that they will be called upon to bear all the costs of environmental protection. Thus, the battle lines are drawn. Property rights, freedom, due process, just compensation, pollution, crowding, and vanishing ecological habitats are all matters of apparently widespread and legitimate interest.

Each of these concerns relates to the interconnected nature of economic and other activities. According to a prominent representative of this view:

> Each of these separate views of the environment is only a narrow slice through the complex whole. While each can illuminate some features of the whole system, the picture it yields is necessarily false to a degree. For in looking at one set of relationships we inevitably

ignore a good deal of the rest; yet in the real world everything
in the environment is connected to everything else (Commoner,
1971, p. 23).

One need not be an economist to recognize that the social/political/economic systems are similarly interdependent; in the elementary sense, society is a general equilibrium system in which "everything depends on everything else."

It is logical that those concerned with the environment would have turned to government--their least costly alternative--as the means for either correcting or avoiding environmental "abuses," or both. If the unrestrained exercise of private resource entitlements via the market creates undesirable spillovers, those who believe they are injured could, in theory, use the contract process, purchase property use rights, resort to force, or enlist the government in their cause. In the contemporary scene, political activity via government regulation is the least costly of the several alternatives. Also understandable is the choice of land use controls as the preventative and enforcement device. Controlling land use provides authorities with absolute veto power over proposed economic activities as well as with the power to set terms and conditions for landowners or land users.

Contributors to the debate have offered opposing views and interpretations. In a study sponsored by the Council on Environmental Quality, authors Bosselman, Callies, and Banta (1973) write as follows:

> A solution must make economic sense, have political acceptability, avoid harmful side effects, allow efficient administration and on and on. Solutions to environmental problems are like chains with many interconnected links. The taking issue is the weak link in many of these chains. All over the country attempts to solve environmental problems through land use regulation are threatened by the fear that they will be challenged in court as an unconstitutional taking of property without compensation. We do not claim that the strengthening of this one link is a quick cure for all environmental issues. Land use regulation is only one of many tools, suitable for some but not all environmental problems, and the chain of land use regulation has many other links, constitutional and otherwise. Nevertheless, if the challenge posed by the taking issue can be overcome we believe it will make a very significant impact on environmental quality (pp. iv-v).

The Supreme Court could overrule Pennsylvania Coal and return to the strict construction of the taking clause by declaring that a regulation of the use of land, if reasonably related to a valid

public purpose, can never constitute a taking. A dramatic gesture
of this magnitude would go a long way toward overcoming the popular myth that land can't be severely reduced in value through regulation. It is this myth which has led so many people to shy away
from undertaking needed regulatory measures (Ibid., p. 238).

THE POLITICAL ECONOMY OF REGULATION AND TAKING

If we confine our discussion of the taking issue to positive economics (what *is* rather than what ought to be), we must conclude that regulation without compensation is the rule rather than the exception in recent experience. In particular, virtually all of the so-called environmental land use controls of the last two decades have been promulgated as regulations, not as takings. Generally speaking, government regulates; it does not take with compensation. Federal, state, and local environmental ordinances, general plans, overlays, and so forth, have become increasingly numerous and constraining in recent years. If anything, they promise to evolve from passive to active in the near future. That is, land use regulation in the past has been passive to the extent that the government responds yea or nay when property owners initiate requests for permission to change or develop their property. Authorities may approve, but usually reject or negotiate an alternative that better suits their purposes. Given current trends, one can forecast that land use regulators will shortly have the power to initiate plans and compel activities of their choosing on private lands.

Many states now exercise such authority with respect to private forest lands; for example, private timberland owners may be directed to reforest their lands after cutting. If they do not perform as instructed, the state will reforest the land and send the bill to the owners. The trend is clear, and it is increasingly only a matter of how bold the agency wishes to be. Government agencies may soon instruct private landowners on what species to plant, where, when and how to plant, and when and how to harvest. All of this is coming, if it is not already here. In addition, authorities will also decide whether or not timber can be harvested (the subject is discussed later in this chapter).

Although the taking issue may seem remotely philosophical and even academic to many, the issue does have a number of fundamental dimensions. Let me explain briefly what I do not intend to address in the following sections. I am not going to attempt to explain why the courts have allowed private property rights to erode. Nor will I make the economist's usual arguments that the real issue (and its solution) is one of efficiency and cost-benefit ratios.

Why not? After all, there is persuasive evidence that in an uncertain world where no one is quite sure what the future holds, a competitive private property system promotes efficiency and consumer welfare better than any alternative yet know (see, for example, Johnson, 1977). Indeed members of the economics profession invariably use efficiency criteria when they argue policy issues. Yet, if my assessment of the record of past battles is correct, efficiency arguments rarely carry the day. I explain my decision not to elaborate the point because I don't believe, as a practical matter, that politicians, lawyers, judges, environmentalists, and the general public are really concerned with efficiency. It is difficult to recall a single instance when our political system at any level has chosen the most efficient alternative from the set of those available.

Which approach to the taking issue remains? Political economy, complete with all of its symbolism and its emphasis on equity and redistribution. In particular, who gets hurt and who gets helped by land use regulations?

I'd like to set the stage for further discussion by suggesting an analogy between the kind of justice that prevailed at frontier town trials in old western movies and that of public hearings before zoning boards, city councils, and boards of supervisors at which landowners attempt to secure permission to exercise their property rights. We are all familiar with the former--a drifter is accused of cattle rustling, is tried before a jury of townspeople and local witnesses for the prosecution (his own character witnesses are too far off to testify), and then is hanged, legally, of course (an illegal hanging was barely averted the day before by the town marshal's exhortations on law and order). The example may sound strident, but I submit that landowners at public hearings receive the same brand of fairness, impartiality, justice, and observance of due process as the drifter did. Why? Because such hearings provide the opportunity for wealth transfers from the politically impotent few to the politically active many, without the constraint of constitutional protection of individual property rights.

FOREST PRACTICE REGULATIONS

The owners of private forest lands are subject to a variety of regulations governing stand establishment, cultivation, and harvest practices. The extent and severity of these regulations depend on the legislation, rules, and ordinances promulgated by the several states and their departments of forestry and by local governments. The constitutionality of regulations governing forest practices on private lands has not been at issue since the Dexter case heard by the Washington State Supreme Court in 1949. The U.S. Supreme Court upheld the Washington State decision without comment. That decision established that "the State Reforestation Act, having been enacted in the exercise of the police power for the general welfare in conserving the resources of the State, is not unconstitutional as impairing the obligation of contracts or as taking property without due process of law" (State of Washington v. Dexter).

Even though constitutional, these regulations, by definition, coerce the owners of private forest lands into practices (and expenses) that they would not voluntarily choose. Assuming that private landowners conduct their business with an eye for increasing their wealth, these regulations must necessarily constrain and reduce landowners' wealth relative to what it would be in the absence of the regulations and rules.

Regulations require that the landowner follow "good forestry practices," such practices being defined and dictated by the state authorities. This may mean regulation of practices that contribute to or fail to correct spillovers such as stream pollution, pests, disease, erosion, and fire. The interconnectedness of the activities of separate forest owners is undeniable in this context; however, the mere existence of a potential externality or spillover does not require acceptance of the heroic assumption that the existing regulations are operating at the appropriate scale or that the costs and benefits of such programs are appropriately distributed among the landowners.

By intent or by chance, California's Forest Practice Act of 1945 addressed the economic issues of spillovers tolerably well. The state Board of Forestry and the District Forest Practice Committees were staffed (or controlled, if you like) by members of the industry. Such an arrangement would seem to be ideally suited to the task if the goal was really to avoid externalities and free riding; clearly, the transactions costs were minimized and, in theory, participants in the forest products industry could internalize the costs of the spillovers pertaining to fire, pest, disease, watershed, erosion, and other causes.

However, it would be naive to think that the popular will was really concerned with externalities of timber harvest. By the late sixties and early seventies, segments of the public realized that the political and legal processes could be used to simply stop timber harvest. When the Bayside Timber Company obtained a permit to log its land under the California Forest Practice Act in 1969, some local residents objected and appealed to the county board of supervisors. The board denied permits to the company to log and to build roads connecting the logging site to the state highway. Having already acquired road-building permits from the California Division of Highways, the company initiated legal action and won a decision that the California Forest Practice Act preempted the county's logging permit system. Unfortunately, in 1971 the courts declared the 1945 statute unconstitutional on the grounds that the regulations were being promulgated by parties having a pecuniary interest in the outcome (Bayside Timber Co. v. Board of Supervisors of San Mateo County).

As a result, effective control of the Board of Forestry is now in the hands of public members who are precluded from having a significant financial interest in the health of the forest products industry. This occurred with the passage of the Z, berg-Nejedly Forest Practice Act of 1973 (California Public Resources Code, 1975). The Board of Forestry was expanded to nine members, five of whom were chosen from the "general public," three from the forest products industry, and one from the range and livestock industry. Thus, the majority of the board has no personal incentive to internalize the real spillovers--other than their desire to do good works. Instead, they have the incentive to foster, promote, and minister to the public's "need" for natural vistas and wildlife habitats, rather than timber production.

More than 90 percent of Californians are urban, already have their shelter, and are only trivially concerned with timber output and the plight of forest landowners. Only 2 percent of California's population resides in its seventeen timber-producing counties. And why should the 98 percent be concerned with timber output? "I think that I shall never see a poem lovely as a tree" gets little competition from "I truly think my spirit soars when contemplating two-by-fours." And we all know that those with the best poems, the best posters, the best songs, and the best Hollywood endorsements are, in contemporary American politics, bound to win.

Most urbanites seem fascinated with the concept of natural vistas and wildlife habitats, to say nothing of trees standing in majestic splendor and having the right to be left that way so future generations have the option of witnessing their demise from natural causes. The emphasis is on future generations regardless of the cost to present landowners. According to Charles Warren, former chairman of the California State Assembly Land Use, Resources, and Planning Committee,

> The process whereby government has the prerogative to regulate private land activities is well established in law, and politics defines the balance between present and future values; but up to now the balance has been leaning in the wrong direction.
>
> The problem lies in acceptably defining the public interest in private property. In the forest management sector, the Forest Practice Act, passed by the California Legislature in 1973 attempted to state this interest. The law established that future productivity of some on-site values, namely the soil and timber resource, is not truly private property in a traditional sense but rather a public property and thus the responsibility of government to protect. Therefore, the state has found it necessary to constrain some present use of these natural resources in deference to their future needs (Warren, 1971, p. 1).

Rarely do we find so clear and bold a statement of the rights and duties of government relative to the private property owner: "Government has the prerogative to regulate private land activities" and "politics defines the balance [point]...." Thus, the message is clear: control the machinery of government via the political process and you control the use of land, private property rights notwithstanding. Indeed, private property rights are an irrelevant issue since timber is public property.

To the extent that regulations do not address spillovers, they are certainly superfluous, since the private timber grower already has intense incentives to follow "good" practices--practices that increase the value (or rate of return) on the woodlot per se. Access to information would not seem to be a problem. Innovative practices are readily observable and good news travels fast--even in the forest. Thus, it is very difficult to make a case for state regulations on grounds that private markets do not provide sufficiently good information to participants. On the contrary, there would appear to be a clear danger that regulations of this variety are inherently ill-suited to the specifics of individual sites. As a consequence, the fine-tuning of techniques is difficult and expensive. In addition, the burden of the regulation falls unevenly on different owners, depending on the physical characteristics of their sites. This is particularly true when such regulations vary among adjacent states, regions, or counties.

Other state regulations more relevant to the issue of the taking of private lands via regulation include the requirement in many states that private forest lands be reseeded within a specified time period after harvest of the previously standing timber. California has such a restocking provision in its Forest Practice Act. Timberland owners must restock their parcels within five years of harvest whether or not they judge that restocking is a prudent investment. Some harvested lands regenerate themselves naturally, but a significant proportion do not.

Other regulations enable the regulatory authority to postpone or forbid the harvest on particular sites. An example is the Douglas County, Oregon Comprehensive Plan. Article 31 of that plan (see appendix to this chapter) defines the concept of a "sensitive environmental overlay." When it is thought that a natural area or resource would be adversely affected in the event of development, that area is protected by a "sensitive environmental overlay" under which development is prohibited and the "primary uses of the district shall be for the conservation of the listed resources." Section 31.010(2) says: "For the purposes of Section 31.000(4) and (5) only, development, as defined by Section 1.090 of the Ordinance, shall include the harvesting of forest products."[2] Section 31.000(4), in turn, refers to "ecologically or scientifically significant natural areas" and 31.000(5) to federal and state wild and scenic waterways and trails.

My translation of the preceding is that if the county declares an area to be "ecologically or scientifically significant" (and just try to find one that isn't), there will be no development in that area. The county also defines the "harvesting of forest products" as a development activity, which means, for the Georgia Pacific Corporation, no timber harvesting on 9,600 acres of its privately owned land in Douglas County; those 9,600 acres constitute a "sensitive wildlife habitat." Will Georgia

---

[2] Article 31 was recently amended to add section 31.000(5) to section 31.010(2).

Pacific be compensated for the loss of the opportunity to harvest timber on its land? No mention is made of compensation, and the reasonable conclusion is that it will not be. The 91,200 shareholders of Georgia Pacific--rich and poor, employed and retired, widows, orphans, insurance and pension fund beneficiaries from all over the country--will pay the full cost of preserving Douglas County, Oregon's ecologically or scientifically significant natural areas, scenic water ways, and recreational trails.

EMPIRICAL INVESTIGATION

Theory alone tells us that regulations such as those described will have an adverse effect on forest product company share prices and, thus, on stockholders' wealth positions. But how significant is the effect?

Data on share prices were collected from the February 6, 1981, issue of The Value Line Investment Survey, a publication containing information on 1,700 publicly traded firms (Arnold Bernhard and Co., 1981). Value Line lists twenty-eight firms in its paper and forest products industry category with stock price data on an annual basis from 1965 to 1980. A dateset of annual share prices by company was constructed by taking the midpoint of the annual range of share price observations for each firm.

In the course of a preliminary investigation, an index of forest and paper company stocks was created as follows: assume one purchased a portfolio of one share of each of the twenty-eight stocks in 1965. The market value of this twenty-eight-stock portfolio was computed for each of the subsequent years through 1980. The annual portfolio value was divided by the Standard & Poor's 500 stock average to control for marketwide influences. The resulting index of real value for the portfolio is plotted in figure 5-1 with the 1965 index value equal to 100. The index suggests the composite portfolio of paper and forest products stocks performed somewhat better than the entire stock market.

However, this time period was characterized by lumber prices that rose faster than the Wholesale Price Index for all producer goods. Figure 5-2 shows the separate time series for all Douglas fir and Southern pine prices, deflated by the Wholesale Price Index with 1966 as a base year. Consequently, the behavior of product prices might suggest that the twenty-eight-firm stock price should have performed even better than figure 5-1

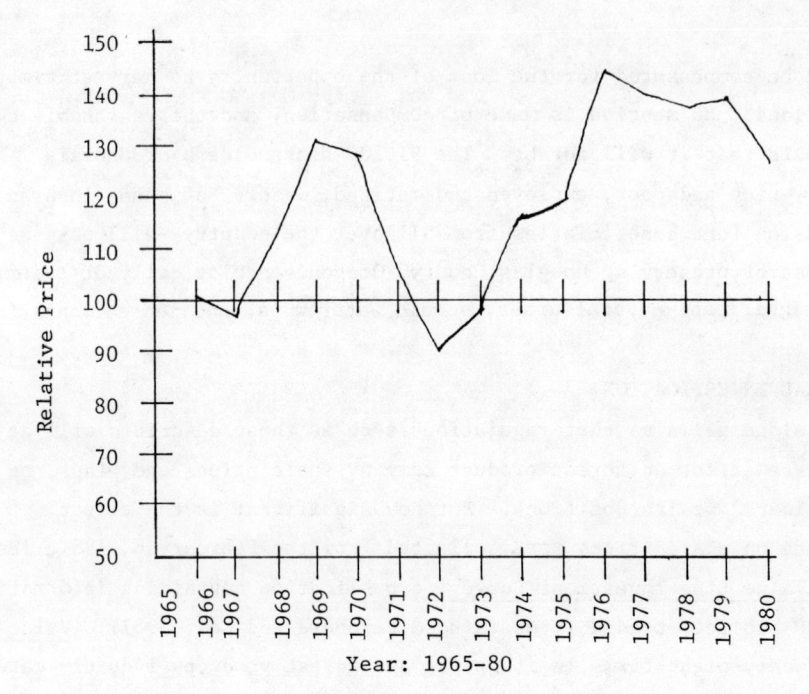

Figure 5-1. Index (forest prices/S&P index) of real value for hypothetical portfolio of twenty-eight forest and paper company stocks.

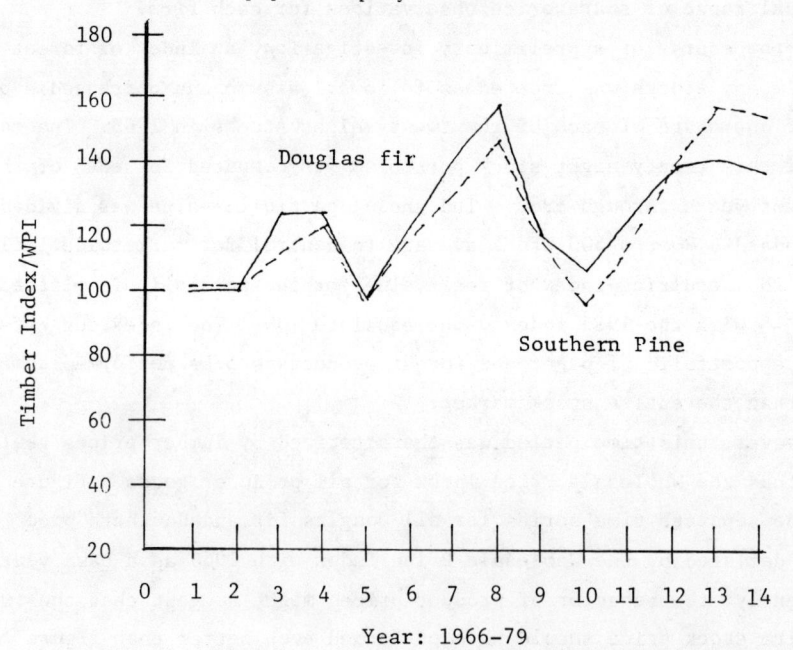

Figure 5-2. Time series for Douglas fir and Southern pine prices (deflated by Wholesale Price Index), 1966-79.

indicates. In order to pursue this point, a somewhat more sophisticated investigation was undertaken.

Standard capital theory establishes that the current market value of any asset depends upon the discounted flow of net expected future income from the asset. Included in the formula are expected discount rates relevant to future time periods and expected rates of growth (if any) of the stream of net income receipts.

In principle, the price of an acre of forest land should also be subject to explanation by this formula. However, forest land prices are difficult to isolate, since different vintage forests have their bare land values commingled with the value of the inventories of timber carried on them. Therefore, I examined the behavior of the stock prices of publicly traded forest products companies during the time period when property rights in timber lands were eroding as a result of regulation. In addition to the Value Line data referenced above, twenty-eight firm dummies, thirteen year dummies, and several variables that give annual price information for specific forest products were also created.

The estimating equation was constructed as follows: Let $S(i,t) = SP(i,t)/M(t)$, where $SP(i,t)$ is the stock price of company i in year t and $M(t)$ is the Standard & Poor's (S&P) 500 stock price index in year t. The dependent variable $S(i,t)$ is defined as the company's stock price deflated by the S&P index to control for factors affecting the stock market as a whole.

Let the real price of forest products, $P(t)$ be defined as $P(t) = FP(t)/WPI(t)$, the nominal price of timber and forest products in year t divided by the U.S. Wholesale Price Index for the corresponding year. This price ratio is common to all forest products firms for a given year. Let $Q(i)$ be the level of net output of firm i. This quantity embodies differences in the product mix and all other firm-specific differences in net output. Capital asset theory predicts that the current price of a firm's stock will depend on the discounted present value of the firm's net output over the lifetime of the firm, so that if $P(t)$ is expected (in year t) to grow at a rate $g(t)$ and $Q(i)$ is expected (in year t) to grow at rate $h(t)$, then

$$S(i,t) = kP(t)Q(i)\sum_{j}[(1 + g(t) + h(t))/(1 + r(t))]^{j} \qquad (1)$$

where r(t) is the market discount rate prevailing in year t and $j = 1, 2, \ldots, n$ with n the projected lifetime of the firm. The constant k is needed to reconcile the units on both sides, since the left side is a ratio of two stock price terms rather than a dollar value.

Taking the natural logarithm of both sides of (1) gives:

$$\ln[S(i,t)] = \ln(k) + \ln[P(t)] + \ln[Q(i)] + \ln[\text{sum}[(1 + g(t) + h(t))/(1 + r(t))]\exp(j)] \quad (2)$$

Since P(t) on the right-hand side of equation (2) is common to all firms, it can be transferred to the left-hand side of equation (2):

$$\ln[S(i,t)/P(t)] = \ln(k) + \ln[Q(i)] + \ln[\underset{j}{\text{sum}}[(1 + g(t) + h(t))/(1 + r(t))]\exp(j)] \quad (3)$$

The Q(i) are firm-specific and not observed, and the expectations/discount terms involving g(t), h(t), and r(t) are specific to each year and also not observed. However, given observations for each firm's stock price in each year, the Q(i) and the expectations/discount terms can be estimated by regression analysis. Introducing dummy variables for each firm (call these F(i)) and for each year (call these Y(t)), the appropriate estimating equation becomes:

$$\ln[S(i,t)/P(t)] = C + \underset{i}{\text{sum}} a(i)F(i) + \underset{t}{\text{sum}} e(t)Y(t) + u(i,t) \quad (4)$$

where C is the constant, and u(i,t) is the disturbance term. In this equation, the coefficients a(i) of the firm dummy variables are the estimated values of the Q(i), while the coefficients e(t) of the year dummies are the estimated values of $\ln[\text{sum}[(1 + g(t) + h(t))/(1 + r(t))]\exp(j)]$. The estimated coefficients for the firm dummies will capture those characteristics (such as output mix, and so forth) that are unique to each firm, while the coefficients for each of the year dummies should reveal changes over time in the industry's price expectations and discount rates. If M firms are observed over T years, there will be M x T observations and hence sufficient degrees of freedom to estimate the parameters of (4).

One technical feature of the estimation should be mentioned here. Since the estimating equation includes a constant term, one of the firm dummies and one of the year dummies must be omitted to avoid perfect multicollinearity. Then the coefficients of the included dummies will represent the differences between the coefficients of the included variables and the corresponding omitted dummy. Thus, the coefficient of F(2) is actually the estimate of the difference a(2) - a(1). In the results reported below, F(1) (corresponding to APL Corporation) and Y (1966) are the omitted dummy variables, so this company and year can be thought of as the bases against which the other firms and years are compared.

Examination of the results presented in table 5-1 reveals several interesting features. Most of the estimated coefficients of the firm dummies are statistically significant at the 5 percent level or better, and these coefficients vary considerably in magnitude. This indicates that there are substantial firm-specific variations in either outputs or costs or both, which is consistent with the varied characteristics of the firms in the sample. (Note that the Q(i) can also capture differences in per share stock values attributable simply to differences in the number of shares outstanding.) The primary focus of interest here, however, is on the estimates of the coefficients of the year dummies, since these are the indicators of how the firms' views of the future were changing over time. Table 5-1 shows that in many years, the estimate of e(t) is not statistically different from the view prevailing as of 1966. Beginning in 1971 and continuing through 1974, however, the estimates indicate a decline in the e(t) values from their 1966 level. In 1972 and 1973, the differences e(t) - e(1966) are statistically signficiant at the .0001 level. Since the e(t) are logarithms of the expectations/discount factors, the 1973 coefficient of -.467, for example, indicates that

$$[\text{sum}[(1 + g(1973) + h(1973))/(1 + r(1973))]\exp(j)]/[\text{sum}[(1 + g(1966) + h(1966))/(1 + r(1966))]\exp(j)] = .63 \qquad (5)$$

Now, a decline in the e(t) can come about either because of a decline in the expected rate of growth of real net output prices, a decline in the expected rate of growth of net output, or because of an increase in the discount rate. For purposes of the present short-term analysis, it is appropriate to assume that the real discount rate r(t) remains roughly constant. From 1970 through 1973, real timber prices in the United States were increasing (see figure 5-1), so it is unlikely that the statistically significant decline in the e(t)'s of 1972 and 1973 could have come about

Table 5-1. Index for 28 Firm Dummies and 13 Year Dummies Using Douglas Fir Prices

Variable		Parameter estimate	Standard error	T Ratio	Prob>:T:	Variable label
INTERCEPT	1	-2.465955	0.107643	-22.9086	0.0001	
DUM2	1	-0.040101	0.124745	-0.3215	0.7481	BOHEMIA INC.
DUM3	1	1.068362	0.119632	8.9304	0.0001	BOISE CASCADE
DUM4	1	0.871565	0.119632	7.2854	0.0001	CHAMPION INTL.
DUM5	1	0.288117	0.119632	2.4084	0.0166	CHESAPEAKE CORP. VA.
DUM6	1	-0.074398	0.121995	-0.6098	0.5424	CONSOLIDATED PAPERS
DUM7	1	1.245036	0.119632	10.4072	0.0001	CROWN ZELLERBACH
DUM8	1	0.523616	0.119632	4.3769	0.0001	DOMTAR INC.
DUM9	1	0.263882	0.135862	1.9423	0.0530	FORT HOWARD PAPER
DUM10	1	0.811112	0.119632	6.7801	0.0001	GEORGIA PACIFIC
DUM11	1	0.786002	0.119632	6.5702	0.0001	GREAT NORTHERN NEKOOSA
DUM12	1	0.775169	0.119632	6.4796	0.0001	HAMMERMILL PAPER
DUM13	1	1.421981	0.119632	11.8863	0.0001	INTERNATIONAL PAPER
DUM14	1	-0.339080	0.147519	-2.2986	0.0222	JAMES RIVER CORP.
DUM15	1	1.268471	0.119632	10.6031	0.0001	KIMBERLY-CLARK CORP.
DUM16	1	0.313801	0.147519	2.1272	0.0342	LOUISIANA PACIFIC
DUM17	1	0.434240	0.119632	3.6298	0.0003	MEAD CORP.
DUM18	1	-0.035243	0.131532	-0.2679	0.7889	PACIFIC LUMBER
DUM19	1	-0.850696	0.124745	-6.8195	0.0001	PENTAIR INC.
DUM20	1	-0.294225	0.119632	-2.4594	0.0144	POPE & TALBOT
DUM21	1	0.635115	0.119632	5.3089	0.0001	POTLATCH CORP.
DUM22	1	1.034906	0.119632	8.6507	0.0001	SCOTT PAPER
DUM23	1	0.736127	0.119632	6.1533	0.0001	SOUTHWEST FOREST
DUM24	1	0.218450	0.119632	1.8260	0.0688	ST. REGIS PAPER
DUM25	1	1.072666	0.119632	8.9664	0.0001	UNION CAMP CORP.
DUM26	1	0.691342	0.119632	5.7789	0.0001	WESTVACO CORP.
DUM27	1	0.896559	0.119632	7.4943	0.0001	WEYERHAEUSER CO.
DUM28	1	0.701240	0.124745	5.6214	0.0001	WILLAMETTE INDS.
YEAR67	1	-0.0096064	0.097679	-0.0983	0.9217	DUMMY FOR 1967
YEAR68	1	-0.077289	0.094840	-0.8149	0.4157	DUMMY FOR 1968
YEAR69	1	0.072456	0.094840	0.7640	0.4454	DUMMY FOR 1969
YEAR70	1	0.226051	0.094044	2.4037	0.0168	DUMMY FOR 1970
YEAR71	1	-0.141279	0.093313	-1.5140	0.1310	DUMMY FOR 1971
YEAR72	1	-0.414126	0.093313	-4.4380	0.0001	DUMMY FOR 1972
YEAR73	1	-0.467801	0.092041	-5.0825	0.0001	DUMMY FOR 1973
YEAR74	1	-0.029632	0.092041	-0.3219	0.7477	DUMMY FOR 1974
YEAR75	1	0.098581	0.092041	1.0711	0.2850	DUMMY FOR 1975
YEAR76	1	0.155281	0.092041	1.6871	0.0926	DUMMY FOR 1976
YEAR77	1	0.046373	0.092041	0.5038	0.6147	DUMMY FOR 1977
YEAR78	1	0.147441	0.092041	1.6019	0.1102	DUMMY FOR 1978
YEAR79	1	0.205160	0.092866	2.2092	0.0279	DUMMY FOR 1979

Note: SSE 32.158626 F Ratio 29.28
DFE 321 Prob>F 0.0001
MSE 0.100183 R-Square 0.7849

because of a decline in g(t), that is, because of pessimistic expectations
about the future course of timber prices. The remaining possibility is a
decline in the expected rate of growth of net output, h(t).

During the early 1970s, several states enacted regulatory laws that
"strengthened" their forest practice acts and reduced the probability that
firms would be able to harvest their timber inventories. Oregon changed
its act in 1971, California in 1973, and Washington in 1974. It is quite
plausible that the adoption of these laws lowered the expected rate of
growth of output of many timber firms. This decline in h(t) in 1971 to
1974 could account for the low e(t) values in those years, and for the
statistically significant declines in the values of the e(t)'s in 1972 and
1973.

The e(t) values recovered after their 1972-73 plunge. In the sample
period to date, however, only e(1979) is significantly greater than e(1966).
What can account for the mild recovery of expectations? One possibility is
that the expectation of future price increases counterbalanced the adverse
effect of the regulations on expected output growth rates. Future prices
might be expected to rise either because of growth in demand or reduction
in supply; a more detailed structural model of the industry would be required to identify which source was predominant. Ironically, the regulations themselves may have generated expected future price increases after a
lag, so that after the initial drop, the net effect of the regulations on
stock prices was minimal.

These results, while not definitive, suggest that the regulations of
the early 1970s had a significant impact on expectations within the forest
products industry. However, more disaggregated data on prices, costs, and
outputs would be required to enable a model of this type to isolate and
quantify the complete set of factors determining the path of the firms'
stock prices over time. It would be particularly difficult to pinpoint
the exact cause of changes in the e(t) terms, since expectations (the g(t)
and h(t)) are not observed directly. Nevertheless, the results presented
here are indicative of the need for more research in this area, in order
to refine the estimates and enable calculation of the net welfare effects
of the regulations.

Appendix 5-A

DOUGLAS COUNTY, OREGON, COMPREHENSIVE PLAN

ARTICLE 31
SENSITIVE ENVIRONMENTAL OVERLAY

Section 31.000 Application

A "sensitive environment" overlay may be applied to any area where development could have an adverse impact on a resource or natural area. The zone shall be applied to the following areas, as designated by the Comprehensive Plan:

1. Mineral and aggregate resources;

2. Energy resources;

3. Fish and wildlife habitats and areas;

4. Ecologically or scientifically significant natural areas;

5. Federally designated wilderness areas or wild and scenic waterways; State of Oregon recreation trails or scenic waterways; however, only to the extent specifically provided in the policies of the Plan.

Section 31.010 Uses Permitted

1. Uses permitted in the areas designated by Section 31.000(1-5) shall be as set forth in the underlying zoning district; however, the primary uses of the district shall be for the conservation of the listed resources.

2. For the purposes of Section 31.000(4) only, development, as defined by Section 1.090 of the Ordinance, shall include the harvesting of forest products.

Section 31.020 General Criteria for Approval

The following criteria shall be applied to development applications in this district:

1. The Approving Authority shall determine by way of Administrative Action, on the basis of findings and conclusions from the application and the record, the nature of the conflict between the proposed development and the identified resource; the economic, social, environmental, and

energy consequences of the proposed development; and what programs or conditions shall be developed to insure that the applicable sensitive environment is protected.

2. The Approving Authority may require further detailed information including site plans and data from the applicant to determine compliance with plan policies and conditions necessary to do so. The Board may prescribe by resolution requirements for information to be submitted with respect to particular designations.

Section 31.030   Pre-Application Conference

1. At the pre-application conference, the Director shall provide the applicant with an opportunity to explain his or its development concepts, and for the Director to explain to the applicant the criteria for development approval when this chapter and the policies, ordinances, standards, opportunities and constraints are applicable to the site and to the type of development proposed, before the applicant has invested substantial money or effort in development design or is committed to a particular type of development.

2. After the required pre-application conference, the Director shall transmit to the applicant, and place in the application file, a memorandum of the minutes of the conference and shall set forth the specific requirements for approval of the application.

3. The applicant may waive in writing the request of a pre-application conference and the memorandum provided by subsection 2 of this Section.

REFERENCES

Arnold Bernhard and Company. 1981. <u>The Value Line Investment Survey</u>. (February 6) (New York).

<u>Bayside Timber Company</u> v. <u>Board of Supervisors of San Mateo County</u>. 20 Cal. App. 3d 1, 97 Cal. Rptr. 431, 3 ERC 1078 (1st Dist. 1971).

Bosselman, Fred, David Callies, and John Banta. 1973. <u>The Taking Issue: An Analysis of the Constitutional Limits of Land Use Control</u>. Council on Environmental Quality. (Washington, D.C.: Government Printing Office).

California Public Resources Code. 1975. ##4511-628 (West. Supp.)

Commoner, Barry. 1971. <u>The Closing Circle</u> (New York, Knopf).

Johnson, M. B. 1977. "Planning Without Prices," <u>in</u> Bernard H. Siegan, ed., <u>Planning Without Prices</u> (Lexington, Mass., Lexington Books, D. C. Heath).

Siegan, Bernard H., ed. 1977. <u>Planning Wihtout Prices</u> (Lexington, Mass., Lexington Books, D. C. Heath).

<u>State of Washington</u> v. <u>Dexter</u>. 32 Wash. 2d 551, 202 P.2d 906. 1949.

Warren, Charles. 1976. Speech before Northern California Section of the Society of American Foresters, Sacramento.

Fred P. Bosselman

COMMENTS

In commenting on chapter 5, I do not want to imply that I have any personal familiarity with California forest practice regulations. Therefore, I must deal with the issues on a somewhat abstract level.

Bruce Johnson says that many of these forest practice regulations are economically inefficient, are based on governmental purposes of secondary importance, and reduce the profit-making capabilities of forest landowners.

While I cannot speak to the particular California regulations, I know of many state and local regulations that affect the use of land in various ways. Many of these regulations are economically inefficient, many are based on governmental purposes of secondary importance, and many have adverse impact on property's profitability. Some are absolutely outrageous and would strongly encourage me to vote against the elected officials who authorized them.

My comments, however, will focus not on whether such regulations are good or bad, but on whether regulations having these characteristics can be attacked in the courts as being in violation of the taking clause of the Constitution. In this regard I would like to comment on two issues raised in chapter 5. First, is there a trend toward increasing governmental interference with landowner's property rights? Second, have the courts proved unwilling to interfere when regulations become severe?

It is a common assumption that America has degenerated from a blissfully unregulated state of nature in the days of the Founding Fathers to our present morass in which a bureaucrat is found behind every tree. In fact, however, the Founding Fathers were subject to a wide range of strict land use regulations. Pennyslvania is a good example.

One of Penn's original "Conditions and Concessions" in 1681 required landowners to leave one acre of trees for every five acres of land cleared (1 Stat. 469). An attempt was made especially to preserve oak and mulberry trees for silk manufacturing and shipbuilding. This taking was a considerable one for farmers, forced to sacrifice a fifth of their productive land; yet no compensation was available. In the city, shady trees were required by a 1700 law. In "An act for regulating streets," it was declared that

> [E]very owner or inhabitant of any and every house in Philadelphia, Newcastle and Chester shall plant one or more tree or trees, viz., pines, unbearing mulberries, water poplars, lime or other shaddy and wholesome trees before the door of his, her or their house and houses, not exceeding eight feet from the front of the house, and preserve the same, to the end that the said town may be well shaded from the violence of the sun in the heat of summer and thoroughly be rendered more healthy (2 Stat. 66, ch. 53).

The early Virginians were subject to strict laws governing the planting of crops. An early Virginia statute established a minimum price for tobacco. The statute went on to regulate strictly the nature and the amount of the plantings.

> <u>Secondlie, be it further ordered</u>, That no planter or master of a familie, shall plant or cause to be planted above two thousand plants of tobacco per pol, and that they which shall not plant tobacco or they which shall be otherwise imployed shall not transfer or make over theire right of plantings unto any other. . . .
>
> <u>Thirdlie, It is ordered</u>, That no person or persons shall tend or cause to be tended above 14 leaves, nor gather or cause to be gathered above 9 leaves uppon a plant of tobacco, and the severall comanders shall hereby have power to examine the truth hereof and yf any offend, to punish the servants by whippinge and to binde over the masters unto the next quarter cort at James Citty, to be censured by the Governor and Counsell. And this act shall continue in force untill the first day of March next ensuinge (Laws of Virginia, September, 1632--8th Charles 1st, Act XX, p. 189).

Both in urban and rural areas, strict regulation of land use was commonplace before and after the adoption of the Constitution and the Bill of Rights. Only much later, during the westward expansion of America and the rapid growth of an industrialized society, did America develop its image as an unregulated playground of railroad tycoons and cowboy gunfighters. The more intensive regulation of the twentieth century can therefore be more accurately characterized as a return to America's original condition than as a departure from its original ideals.

I raise this issue not to defend bad regulation, but merely to point out that there is scant evidence that bad regulation of land use was a serious concern of our Founding Fathers or that they sought to contain such regulation through any provision of the Constitution.

We have long passed the point in American history, however, when Supreme Court justices were embarrassed about inventing constitutional rights. President Nixon announced that he was going to appoint strict constructionists who would follow the original intent of the Constitution, but his four appointees no more fit that mold than do their five fellow justices. In the law schools this is known as "an evolving constitution"; that is, the Constitution means what the Supreme Court says it means.

In recent years the Supreme Court has been taking an increased interest in property rights and has held that such rights are protected by the takings clause against certain types of regulation.

The most recent expression of the Court's views on this subject is the opinion of Justice Brennan in San Diego Gas & Electric Co. v. City of San Diego (Docket No. 79-678), decided March 24, 1981. Although Justice Brennan expressed his views in a dissenting opinion, they would appear to represent the views of a majority of the justices.

Justice Brennan described the point at which a land use regulation becomes a taking as the point at which the regulations "destroy the use and enjoyment of private property." He went on to say:

> From the property owner's point of view, it may matter little whether his land is condemned or flooded, or whether it is restricted by regulation to use in its natural state, if the effect in both cases is to deprive him of all beneficial use of it. From the government's point of view, the benefits flowing to the public from preservation of open space through regulation may be equally great as from creating a wildlife refuge through formal condemnation or increasing electricity production through a dam project that floods private property (pp. 16-17).

In summary, the Supreme Court is making it increasingly clear that it is willing to hear challenges against unduly severe land use regulations under the takings clause. To meet the test, however, the regulations must substantially deprive property owners of the use of their property for economically beneficial purposes. The application of that test to individual situations, and the determination of the amount of any damages to which the property owner is entitled, will provide gainful employment for lawyers and economists for many years to come.

PART III

MANAGEMENT OF PUBLIC FOREST LANDS

OVERVIEW OF PART III

The matter of public ownership of land in a private market economy has always generated questions. Within the United States the management of public forest lands continues to be surrounded by considerable controversy. Should large portions of the forest lands be publicly owned? If so, how should they be managed? Do public ownership and management serve societal interests? These questions are examined in the following chapters and comments.

Originally, many of the forests became part of the public domain by default. For the most part these forest lands were never claimed by private parties under the Homestead Act, while a much smaller portion of them was claimed but eventually reverted to the government in lieu of back taxes.

Most of this nation's public forest lands are administered by the Forest Service, with smaller areas being under the jurisdiction of the Bureau of Land Management or the various states. For many years the Forest Service viewed its role as that of a custodian or steward of the public forest lands. In time, however, the public forest lands became more important as a source of timber, and gradually the recognition grew that these lands provided, or were capable of providing, a wide array of other outputs such as erosion control, watershed values, wildlife, recreation, and so forth. In 1960 the Multiple Use Sustained Yield Act authorized the Forest Service to manage for multiple use outputs. The Forest and Rangeland Renewable Resources Planning Act (RPA) of 1974, as amended by the National Forest Management Act (NFMA) of 1976, mandated that the Forest Service undertake periodic comprehensive, systematic planning. It is this planning--the justification, the process, the techniques and the outcome--that are the focus of the next three chapters.

In chapter 6, Krutilla, Bowes, and Wilman provide a justification of public ownership and management of forest lands, an interpretation of the recent legislation, and a discussion of various conceptual and technical difficulties involved in the planning process. The authors examine the role of government, and particularly of the Forest Service, in the economic sphere as it pertains to forest lands. They argue that of the three roles of economic stabilization, resource allocation, and income distribution, a compelling argument can be advanced that the Forest Service should involve itself only in the work of the allocative branch of government.

Krutilla, Bowes, and Wilman go on to argue that "the predominant thrust of the legislation goes to the issue of allocating scarce forest land and rangeland resources among their various competing uses." They maintain that the principal thrust of recent legislation calls for planning in accordance with the economic maximization of social net returns. The forest is capable of producing a variety of valued outputs. These include watershed protection, wildlife, and recreation, as well as timber values. For various reasons, some of the outputs may be either unpriced or incorrectly priced. Tradition has established that a lower-than-market price be applied to recreational services provided by public lands, while the existence of externalities associated with some outputs, for example, watershed or widely ranging wildlife, may make it difficult for forest land owners to capture the value of their forests' outputs, and thus the private economic incentive to produce these outputs is inconsistent with the social values the outputs provide.

The final section of the chapter sketches a planning approach that draws on the cumulative scientific research of the Forest Service and on the areas of management science and economics. Briefly summarized, the approach views the forest as a multiproduct firm. A production function relates inputs to outputs, with the level and quality of the various outputs being interdependent. The problem of forest management is similar to that of the multiproduct firm: maximize some objective function (profits for the private firm, social value for the public manager) subject to the cost constraints. In practice, of course, this is difficult. Demand curves for the nonmarket outputs must be estimated; production interdependencies are often poorly understood and vary substantially across sites. Then, too, the choice of which alternative management regimes to examine is limited by computational constraints. The authors suggest that preliminary screening to identify promising alternatives might be accomplished with the use of a simulation model.

Given their view of the inherent complexity of the problem, Krutilla, Bowes, and Wilman are quite impressed with progress made thus far by the Forest Service. While recognizing weaknesses in the current procedures and techniques, they nevertheless believe that the Forest Service has progressed remarkably toward making public forest planning operational, and thereby toward providing a more socially desirable mix of forest outputs.

In chapter 7, Douglas Leisz, associate chief of the Forest Service, addresses the question of the impacts of the legislatively mandated RPA/NFMA planning process on management and planning in the Forest Service. Leisz notes that an attempt to establish a clear causal relationship between a legislative provision and a subsequent change in agency behavior is certainly a difficult task. In addition, he believes that it is too early to determine the degree to which the promise of the RPA/NFMA planning process will be realized. Thus, the chapter represents a preliminary assessment of the planning process and its impacts. Leisz lists the major impacts on the Forest Service as follows: first, the RPA and NFMA have led, for the first time in recent history, to strategic and long-term planning for renewable resources on a national scale; second, the process has given policy and program direction a central focus--programs are now formed with national objectives using locally derived data; third, the planning process has accelerated the trend toward a broader mix of professional skills and improved analytic capacity within the Forest Service.

The results of the changes listed above are reflected in an improved analytical approach to planning, an improved budget dialogue, and the provision of a vehicle for program and policy change. A further change already noted, is the shift toward centralization in an agency that had traditionally been decentralized.

Leisz suggests the following midcourse changes in the planning process: first, streamlining the process to make it less complex and less costly; second, distinguishing more carefully the appropriate amount of detail for each planning level; and third, having resource management programs reflect more clearly the economic, social, and cultural realities of the nation.

In chapter 8, John Walker provides an interesting complement and contrast to the material in chapters 6 and 7. Krutilla, Bowes, and Wilman develop the justification for public ownership of forest lands, provide an economic interpretation of the legislative mandate, discuss techniques of public management of some forest lands, and generally praise the Forest Service effort; Leisz discusses the impact of recent legislation, particularly as it affects the Forest Service; and Walker assesses the performance of the Forest Service's planning effort and its management practices, finding both decidedly deficient. Walker's criticisms stem from the central thesis of his chapter, which is that economics does not play a critical role in Forest Service planning. An implication of this is that the management of the national forests deviates substantially from the social optimum. The thesis appears, upon first reflection, to be in conflict with the considerable resources that the Forest Service is directing into analytical and model-building activities that have major economic components. However, Walker argues that noneconomic institutional constraints such as the "allowable cut effect" and "nondeclining even flow," as introduced into the planning process, so constrain the outcomes of the planning tools that the results are dominated by these constraints. Thus the planning tools used and the management regimes actually practiced are decidedly noneconomic in nature. Walker argues that the National Forest Management Act is not a prescriptive act and that, while taken as a whole it provides numerous requirements for economic analysis, it also contains numerous ambiguities and inconsistencies. Thus, the National Forest Management Act gives the Secretary of Agriculture and the Forest Service a great deal of flexibility. From his discussion of the actual workings of the planning process and the resulting management, it is clear that Walker is greatly disappointed in the direction taken and the progress to date, and that he believes the resulting process and management do not serve societal interests.

While Walker's argument and examples relate largely to timber outputs, he indicates that his comments are intended to cover the management of both the timber and nontimber outputs of the forest. He concludes that he is in essential agreement with every point except one made by Krutilla and coauthors. That exception involves the assessment of progress that is being made in moving toward the desired goals. Walker perceives the pace in that direction as "glacial."

In his discussion of the Krutilla, Bowes, and Wilman chapter, Clark Binkley focuses on some of the problems inherent in the planning process. He notes that there are at least two dimensions to the strategy for planning: the choice of analytical techniques and the optimal application of the appropriate techniques. He raises questions about both dimensions. Is linear programming appropriate to the nonlinear system? Do the benefits of current planning exceed the costs? The majority of his comments relate to the cost-benefit issue. Binkley notes that a plan is optimal only with respect to some specified and unknowable future state of the world. He then argues that in a world of such uncertainty, the costs of being wrong must also be considered. The plan must recognize the costs of operating in a world different from the one for which it was designed. However, the deterministic planning process, according to Binkley, ignores adaptive responses to unforeseen events.

Given the costs of planning, the lack of adequate scientific information, and the difficulties of adapting the plan to changing conditions, Binkley makes an innovative proposal: let each forest become a profit center in an accounting sense, charged with maximizing the social net value of its outputs given transfer prices for extramarket resources and assessing realistic costs of inputs. Forest managers would then be judged on the basis of net return. Such a system would provide incentives for individual forest managers to manage to maximize social profitability within the context of a decentralized decision-making system.

In his remarks on chapter 7, Perry Hagenstein comments on the three questions asked by Leisz. In essence, while expressing serious reservations about this country's willingness to move toward national plans, Hagenstein seems on balance to agree with Leisz that the program goals coming out of the planning process have made a difference in budgets and appropriations and therefore have affected national policies for natural resource management. Hagenstein believes that if national forest planning is to help meet long-term resource needs, national plans have to be translated into political support. In conclusion, Hagenstein states his agreement with Leisz that more experience is required before a definitive judgment can be made as to whether budgets and national program decisions are better as a result of the planning effort.

In his comments on Walker's chapter, David Anderson points out that much of the frustration with Forest Service planning is the outcome of time pressures that have forced the Forest Service to develop simultaneously both procedures and plans. Anderson has no quarrel with Walker's ideal of efficient resource allocation. However, he points out that the applications of the planning tools are inherently deficient. Anderson notes that progress is being made. For example, while the estimation of downward sloping demand curves is particularly difficult, an estimation has been attempted in the Rocky Mountain region and elsewhere. However, while agreement has been reached as to the validity of demand curves for timber, the same is not true for grazing and water. But, where agreement is reached, the demand curves are being utilized in the modeling and planning activities.

Anderson agrees with Walker that the constraints should be analyzed, and he maintains that the Forest Service should and does examine alternatives. He argues, in essence, that the economics of the Forest Service planning processes is basically sound and defensible. Finally, he concludes that in the transition to the new planning role, some inefficiencies are necessarily going to be encountered.

Chapter 6

NATIONAL FOREST SYSTEM PLANNING AND MANAGEMENT:
AN ANALYTICAL REVIEW AND SUGGESTED APPROACH

John V. Krutilla, Michael D. Bowes, and Elizabeth A. Wilman

National forests mean different things to different people. To some they reflect a national tradition of making museums, parks, and playgrounds available as "merit goods" (Musgrave, 1959, part I, p. 13) at no expense to users in order to encourage participation in activities that are considered to be in the interest of a civilized society. To others, they represent publicly owned real estate which, in the interest of an overburdened tax-paying public, should be managed to maximize the returns to the U.S. Treasury. In promoting the first view, the statement is often made that since the public owns the national forests, those forests should be viewed as available for public use, not private profit. In advancing the second view, there are dark mutterings about the inequitable burden that users of primary commodities--that is, timber, forage, and minerals--must bear in order to accommodate consumers of final consumption services, namely, the amenities enjoyed <u>in situ</u> by recreationists in the national forests. Both views enjoy some (albeit only tenuous) sanction in American tradition, even though when brought together in addressing the same issue they are found to be in conflict unless managed artfully.

Such opposing views are expressed against a background of eclecticism that grows out of the historical process by which tradition, legal doctrine, and administrative policy have evolved--often piecemeal--to deal

---

[1] Such "merit goods" provide benefits to society in the form of improved mental health or crime reduction. Since these benefits cannot be restricted to the users, they would be underprovided if users supplied themselves or were required to pay the full cost.

with specific situations at particular times in the development of the
nation and its institutions. These may not all be consistent with each
other--indeed are very likely not to be--but have persisted into the
present, providing fertile ground for both excited polemics and serious
discussion. A critical historical analysis investigating the origins of
the existing customs, doctrines, and policies that form the complex of
directives affecting administrative behavior would doubtless reveal much
that is essential and readily justified, but would also reveal much that
is outmoded and counterproductive in the context of today's problems and
management challenges. Such an investigation would be instructive and,
in our view, ought to be undertaken for purposes of general edification.

It is not possible within the space of this chapter, however, to view
the problem of national forest system planning and management historically.
Instead, we will substitute a functional approach in an effort to promote
internally consistent management criteria within the framework of a coherent management philosophy. The chapter proceeds as follows: (1) it
attempts to conceptualize the national forests as public assets managed
to realize goals derived from the theory of the state and political economy; (2) it refers briefly to the legislative provisions that both direct
and constrain the chief of the Forest Service in carrying out his mandate
as the top administrative officer for the national forests; (3) it reviews the Forest Service's interpretation of its legislative mandate by
reference to the regulations issued (pursuant to the Forest and Rangeland
Renewable Resources Planning Act of 1974, as amended by the National
Forest Management Act of 1976), and evaluates, to the extent practicable,
the consistency of its national forest planning with the intent of the
legislation as interpreted for Forest Service personnel through the regulations; and (4) the chapter concludes by sketching an approach which,
while not yet completely operational and thus unavailable for the current
(1980-83) planning efforts on the national forests, nonetheless draws on
the cumulative scientific research and modeling of Forest Service scientists. This approach also draws on the areas of management science and
economics that may provide insight for future planning as well as assistance in developing relevant data for more efficient management than is
possible with the present state of the art and inventory of resources and
research results.

THE ROLE OF A PUBLIC RESOURCE MANAGEMENT
AGENCY IN A PRIVATE ENTERPRISE ECONOMY

For a society with a tradition that is a descendant of the Jeffersonian dictum that the society which is governed least is governed best--a principle reaffirmed by Lincoln's philosophy that government should do for its citizens only what they cannot do for themselves--the following question remains to be considered: why should there be such a large amount of public land held by a government dedicated to the vestment of property rights in private parties and to the production and exchange of wealth by means of market-driven economic processes. On the face of it, there may well be a question, as has been raised in the West, about the justification for holding such vast amounts of land in public ownership.

The historical record shows that much of the public land is, to borrow from the Shands and Healy (1977) title, real estate that nobody wanted at the time. These were the less desirable lands passed up by homesteaders, railroads, mining companies, and all others to whom the public domain was parceled out under various land disposal policies intended to get the country settled and developed; or they were lands that had been abandoned after virgin stands were harvested. But this is not a justification of current public ownership. To find that, we must look to other land withdrawals and reservations made for specific national objectives and at repurchases of lands once privately held--lands repurchased to realize national conservation objectives that were inconsistent with those areas remaining in private ownership (for example, the purchase of private lands to achieve the purposes of the Weeks Act of 1911).

Indeed, public intervention in the affairs of society, even in regard to economic matters, has a long and independent tradition. For example, the year of our Declaration of Independence was also the date of Adam Smith's urgent declaration of independence from the suffocating regulations of a mercantilist (government-dominated) economic system (Smith, 1976, vol. II, bk. IV, chap. IX, pp. 687-688); Smith acknowledged scope for governmental activities in numerous areas like those the first Republican president alluded to. But certainly in America, Jefferson's Secretary of the Treasury Albert Gallatin a half century before Lincoln's time recommended governmental intervention in the interest of improving the infant country's transport system ("Report from the Secretary of the

Treasury, April 6, 1808," 1834). The basic economic rationale for this was more formally developed by Jules Dupuit (1844) in his paper on the benefits of public works in a French document addressed to these issues. From Dupuit through the latter-nineteenth century economists--English, Italian, and Swedish alike--to the advent of twentieth-century Pigouvian welfare economics, the development of political economy, and more recently the development of a theory of public expenditure and justification for public intervention, theories on such intervention followed in lockstep with actual developments at home and abroad in public (or publicly supported private) transportation and in the fields of conservation and resource development.[2]

Thus the growth of a body of thought on the proper role of the government in the affairs of the citizenry has a very long history in political theory and political economy. Indeed, the existence of a private enterprise system depending on the vestment of rights to property in private parties and the exchange of goods and services through a market system in pursuit of self-interest, itself derived from a structure of laws and institutions that have been provided by public intervention of a participating citizenry.

It is commonly held that the institution of private property and the market to organize transactions involving exchanges of goods and services is more efficient, or perhaps less arbitrary (or perhaps less tyrannical), than an alternative; but it is also commonly accepted that the government must intervene for purposes beyond simply establishing the institutions for the free exchange of wealth through individualistic actions mediated through markets. There are generally three areas discussed in the theory of political economy that merit mention before we address the issue of planning and management of public lands and of the national forests in particular.[3] These are discussed briefly below.

Richard Musgrave summarizes the roles of government in the economic sphere, characterizing them as performed by the three branches of economic

---

[2] For a relatively modern synthesis and reformulation see Krutilla and Eckstein, 1958, part I; see also Musgrave, 1959, part I.

[3] For a somewhat more extended discussion see Krutilla and Haigh, 1978, Section II, pp. 375-384.

stabilization; resource allocation; and income, wealth, or welfare distribution. The first of these has little to do with the other two except in that, if successful, it keeps the free market machine in working order. It is primarily concerned with the increasingly difficult business of evening out the fluctuations in employment and price levels that can occur in a free economy, and thus facilitating the operation of the economy and preventing hardship to its citizens. The allocative branch is concerned with the efficiency with which resources are used. Under a given set of preconditions, including sufficient competition and the condition that all relevant effects are reflected through prices, the voluntary exchange of goods and services as mediated through free markets will lead to an efficient use of productive resources, whether primary commodities, intermediate goods and services, or final consumption goods, so that no reallocation could make everybody at least as well off. Such efficiency is necessary for the achievement of maximal social welfare; sufficiency involves, in addition, an acceptance of the resultant distribution of income and wealth. However, the preconditions necessary for such outcomes to be realized through market exchanges do not always exist, for a variety of reasons discussed elsewhere (e.g., Krutilla and Eckstein, 1958; or, more briefly, Krutilla and Haigh, 1978), and in order to provide for an efficient allocation of resource and final consumption goods and services, the government may intervene to assist the existing institutions in achieving a more efficient allocation. It should be noted that although perhaps less fundamental, this intervention is of the same sort as the intervention of government originally in establishing, through the Constitution and subsequent legislation, the institution of private property and the sanctity of the contract--two conditions without which the private market system could not function.

The third area in which it is generally agreed that the government has an obligation to intervene is in the matter of equitable distribution of income, wealth, or, if you will, welfare of its citizens. There are various reasons for this discussed elsewhere. The market allocation of resources is conditional upon the initial distribution of titles to property and wealth. There are different efficient allocations of resources likely for each different distribution of income and wealth. Such initial distributions and resulting allocations may be viewed as sufficiently inequitable to justify intervention. In the extreme, a market may allow accumulation and concentration of wealth in so small a proportion of the population as to risk the basic conditions essential to fostering a free society with its free enterprise system.

Given these various roles justifying public intervention, it is desirable to distinguish among different types of executive branch agencies and the roles for which they are fitted in order to determine what the role of the Forest Service itself, as an agent of the government, might be. The class of agencies conventionally referred to as land and resource management agencies includes the Forest Service and the Bureau of Land Management. (Forest Service responsibilities of course also include research functions and assistance to state and private forests.) Some agencies are concerned with supplying public goods such as transportation and defense. Others are regulatory agencies; those such as the Environmental Protection Agency and the Food and Drug Administration are directed to the protection of the health and welfare of the public, while others such as the Federal Trade Commission, the Department of Justice's Antitrust Division, and related agencies--even the Federal Reserve System's Board of Governors--monitor and mediate the performance of the economy with respect to both resource allocation issues and stabilization at high levels of performance. Similarly, agencies such as the Internal Revenue Service and the Department of Health and Human Services attend to the tax and transfer issues associated with the government's distributional responsibilities. While not exhaustive, this list at least identifies by function some classes of agencies that are presumed to be suitable for performing the three governmental roles in the economic affairs of society.

A review of the functions required of agencies in these different roles suggests that it can be compellingly argued that the Forest Service as a resource management agency is largely restricted in its basic capabilities to taking an active part only in the work of the allocative branch of the government (see Krutilla and Haigh, 1978). Since there is always some possibility of Forest Service allocation decisions resulting in distributional imbalances, it is arguable that in the area of community stability and distributional equity, the agency should play a passive role-- for example, being circumspect about plans and programs that may destabilize conditions or create inequitable burdens for some particular group or community. However, the tremendous complexity of assessing the distributional consequences of national forest policies and the likelihood of such effects washing out in the long run counsels that consideration should be limited to any large distributional imbalances these policies may cause.

Thus, as a general matter, the primary role, and the one to which the Forest Service must actively address its attention, is that of improving the allocation of resources.[4] It is here that inefficiencies occur, because of the factors that restrict the private sector's ability or incentive to provide some beneficial goods and services. For example, the private sector might have difficulty producing certain wildlife populations because of the magnitude and interconnectedness of the ecological system on which their production is dependent. The scale of investment and ownership may simply be too great. The private sector might also have trouble controlling access to the use of these resources because of the fugitive nature of the wildlife. The inability to control access limits the possibilities for imposing fees and prices and so diminishes the incentives for private production and ownership.

Watershed protection and scenic vistas provide further examples of services difficult to market privately because of the potentially large scale of ownership required and the resulting difficulty in controlling access to such resource services.[5] These and other services like them would be underprovided if their production were left entirely to the pri-

---

[4] The role that public land management can play for the maintenance of a centuries-old Spanish American culture in the general region of northern New Mexico may be advanced as an exception to the statement--but it should be recognized as the exception rather than the rule.

[5] The extreme case in terms of difficulty of excluding access is the "public good." The term "public good" is used here in its more narrow and precise definition as a good (or service) which if provided for one individual is provided for all (indivisible, hence not subject to exclusion), and which if used by one individual is not reduced in the amount available for all other individuals. An example is a scenic vista that may be viewed by one individual without reducing the amount available for viewing by any other. This may be contrasted with a private good, such as a particular vantage point for viewing, which has limited capacity: someone occupying the space reduces the amount available for some other potential viewer for that moment in time. The terms "public" and "private" in this context have nothing to do with whether the resource service is publicly or privately provided, but rather have to do with the technical properties of attributes of the commodity or service.

vate sector, and their prices would not accurately reflect economic value. There also is an absence of contingency and futures markets for these services, so that misallocations could occur between time periods, as well as among current goods and services. It is precisely for these reasons that the public may choose to intervene in the management of these resources. Thus, the Forest Service must engage in land use planning as a supplement to the market to achieve the resource allocation efficiency objectives on which its primary justification exists. In doing so it provides a wider range of resource services than would otherwise be produced.

CONSISTENCY OF NATIONAL FOREST LEGISLATION WITH ECONOMICALLY EFFICIENT MANAGEMENT

By no means all of the Forest Service missions involve strictly improving resource allocative efficiency. But the following examination of the basic legislation confirms the growing preeminence of the directives calling for implementation of rational economic decisions as the guiding management philosophy for the bulk of the Forest Service's programs. The early formation of the national forests can perhaps best be explained as a response to inadequately defined ownership rights and to a concern that private timber interests were, as a result, viewed to be myopic in their use of forest lands. The motivation for passage of the Forest Service's Organic Administration Act of 1897 can be traced to the conflict between two views. The public's concern was that private use of existing public lands and of many private lands was resulting in degradation of potential timber supply and water flow. Until that time private interests had had effectively unrestricted commercial use of public land and were eager to protect this position. Watershed degradation and consequent off-site externalities imposed on users of navigable streams were a major rationale for the Weeks Act of 1911 extending national forests to the eastern United States in various mountainous areas previously occupied by private owners.

With the lands reserved as national forests, management principles gradually evolved. The McSweeney-McNary Act of 1928, which established the forest research experiment stations, first emphasized research into economic considerations in management and next called for the determination and promulgation of economic principles that underlie the establishment of best policies for land management and product utilization. More recently, the provisions of the Multiple Use Sustained Yield Act of 1960 stipulated that forests be managed for a range of both nonpriced and marketed resource services, taking account of the changing relative value of such services. This, in lay terms, is precisely what motivates the management of a private, multiproduct enterprise if it intends to survive in a competitive environment. All of the above legislation was responsive to the problem of adverse third-party effects or involuntary (extramarket) exchanges and was implemented in a nonformal manner based on "professional judgment," "conventional conservation wisdom," or in any event, by rather loosely defined criteria.

The phrasing of the Renewable Resources Planning Act (RPA) as amended by the National Forest Management Act (NFMA) tends to be more explicitly economic. The RPA as amended is written in terms of assessing the supply of and demand for the commodities and services of forest lands and of comparing the benefits and costs of programs formulated to meet the economic demand (that is, providing a quantity that will clear the market under the conditions that social costs and benefits are equal at the margin). This is not to suggest either that these concepts are consistently employed in the legislation, or that all the stipulations are more economic than technocratic.[6] But the predominant thrust of the legislation goes to the issue of allocating scarce forest land and rangeland resources among their various competing uses. This is, of course, the essence of the economic problem--or the problem arising out of the need to share among competing claimants resources that are scarce relative to the demand for them. And the degree to which this is done efficiently represents the extent to which the net public benefits from the management of national forests are maximized.

A careful review of the legislation, particularly of the Renewable Resources Planning Act and the National Forest Management Act, leads persuasively to the conclusion that the recent legislation allows, and perhaps even requires, that the national forests be managed as economic assets with

---

[6] For a systematic analysis of the RPA as amended, which addresses interpretations of potentially inconsistent provisions to provide a coherence in the management directives derived from the legislation, see Haigh and Krutilla, 1980.

maximum practicable efficiency. There is little doubt that one can interpret the legislation as supporting a benefit-maximizing objective. The planning and management paradigm therefore can be conceptualized as a benefit objective function to be maximized subject to meeting certain legislatively stipulated side-conditions. These side-conditions provide for unspecified amounts of the nonpriced resource services required by the multiple use legislation but not valued in a monetary metric. They are thus treated as constraints rather than included as variables in the objective function.

We believe it fair to say that the regulations prepared and promulgated pursuant to sections 6(g) and 15 of the RPA as amended are consistent with the above interpretation of the legislation. That is to say, the forest planning that has begun under RPA regulations represents use of a benefit-maximizing criterion function, with some of the nonmarketed forest outputs entered as side-conditions to be met where the outputs themselves cannot be entered as variables in the objective function because of the absence of prices or values expressed in a monetary metric. From a review of actual plans conducted as of the end of 1980 on the national forests under the new RPA regulations, the procedures appear to be quite consistent with the interpretation sketched above. This is not to say that all of the data required for a completely satisfactory performance are available, or that planning team judgments on relative values either implicitly or explicitly enjoy universal concurrence when the team sets the constraints to achieve levels of nonpriced forest products outputs. All that we are suggesting is that if the data reflecting the cost and benefit functions were available, the approach being used would tend to result in management of the national forests as though they were economic assets, the objective of which was, generally speaking, to maximize the net public benefit. This is indirect evidence of the Forest Service's accepting an interpretation of the legislation that has a significant economic thrust, that the regulations in spelling out the guidelines for National Forest System management acknowledge the economic content of the legislative directives, and that the planning is proceeding consistent with the management of national forests as economic assets.

This may appear to be a commonplace observation; but considering the forest-level plans being undertaken, we are not certain that the forest managers, as distinct from the planning teams, necessarily perceive the implications of the legislation in the same light. This may be due to a significant difference in interpreting what is being viewed, or it may be largely a semantic difference: that is, forest managers, descending from silvicultural and other ecological antecedents, may be moved by different impulses than are people managing other economic assets, or it may simply be that different terminology from one profession to another is not recognized as having essentially the same meaning. It may be useful to explore this issue in more detail below.

AN INTERDISCIPLINARY VIEW OF FOREST MANAGEMENT

When a manager with an economic perspective looks at the problem of managing the resources of the national forest efficiently (for the moment take as given the demand for the forest's resource commodities and services) that manager is likely to summarize the problem by identifying the inputs required by a given production technology, by measuring or otherwise determining the outputs, by transforming inputs into units of account by referring to the price of factor services, and by transforming the output units also into a common metric so that inputs and outputs can be compared in terms of costs and benefits expressed in a common unit of account, namely, in monetary terms. The relation between physical units of inputs and outputs fixed by the technologies available is referred to as production function--a shorthand expression for saying that output is a function of (depends on) the inputs employed, given the technology, that is,

$$f(Q_1, Q_2, \ldots, Q_m; X_1, X_2, \ldots, X_n) \leq 0$$

where the Q's represent the outputs, the X's represent the inputs, and f represents the functional relationship linking outputs to use of inputs.[7] Economists addressing forests or problems in forestry often speak precisely in these terms--that is, in terms of the need to specify a production function as a proper step in the proper understanding of forest management (see, for example, Clawson, 1978).

---

[7] This is a general relationship including possibilities for both technically efficient and technically inefficient production. A strict equality would be required if only technically efficient production alternatives were included.

In elementary textbook economic theory, production is generally addressed in the abstract with smooth, continuous production functions; inputs are represented by broad statistical aggregates. This is done in an apparently static framework. Such simple expository devices are not particularly felicitous for the understanding and analysis of real dynamic planning problems. This is not to suggest that economists have ignored production under conditions that are similar to those addressed in forestry. The intertemporal theory of production and investment is well developed.[8] Indeed, by distinguishing each input and output by point of time, intertemporal production decisions can be understood to be analogous to static production choices--although certainly subject to a greater degree of demand uncertainty. Under such a view, the set of all technologically (biologically) possible sequences of inputs and outputs over time is what the economist is representing in shorthand notation by the production function. In practice, when faced with need to provide numerical solutions to real planning problems, we must face the need to provide detailed descriptions of the known production possibilities. Even so, some degree of approximation of our knowledge is likely to be essential. Such approximation may require limiting the number of alternative production sequences represented, accepting some degree of aggregation of inputs and outputs, and perhaps imposing some structure on the set of production possibilities.[9]

---

[8] See, for example, Fisher, 1954; Malinvand, 1953; and Hirshleifer, 1970. A large number of studies of dynamics in agricultural economic problems exist. Ranching has some of the same problems for management that occur in forestry. For example, see Jarvis, 1974; Spreen, McCarl, and White, 1980; and Chan, 1981.

[9] One structure that seems particularly relevant, and which has been of particular interest to economists, is the recursive technology. Such technologies are the basis of capital theory and the techniques of optimal control and dynamic programming. A technology is recursive if the current production possibilities can be represented as a function of current inputs and the existing accumulation of various stocks, that is, capital goods (the forest "condition"), and are otherwise independent of the past. Such structure allows us to deal with a sequence of production possibility sets of greatly reduced dimension. Each involves only the current outputs and inputs, including among the outputs the modified levels of stocks available in the subsequent period (that is, the subsequent forest condition). The reduced dimensionality allows a fuller representation of the production possibilities, and the structure allows for a greater intuitive understanding of investment decisions.

The national forests, given the legislatively stipulated mandate for multiple use, produce a mix of outputs depending on the decision of whether to harvest or retain a given portion of the standing crop. These standing stocks can be held for their accumulating biological growth and price appreciation or harvested to become marketed commodities. Moreover, different forest outputs result from different silvicultural treatments--precommercial thinning, prescribed burning, shelterwood harvests, clear-cuts, and so forth--as well as from their timing, size, and distribution. That is, the in situ condition of the forest stock, its suitability as wildlife habitat, and its visual condition may themselves be valued by consumers. One may say loosely that the mix of forest resource commodities and services is a function of basic land form and character and of the timing and location of vegetative manipulation. Each different treatment results in a different time sequence of the "state of forest organizations" to draw on a forest ecologist's characterization, or in a different sequence of "stand condition classes," as often characterized by silviculturists.

Accordingly, if one approaches a forest manager for a schedule of inputs and resulting outputs, the problem will be answered in a dynamic framework. The elements of forest resource commodity and service production will be characterized in terms of a prescribed combination of scheduled activities. These prescriptions will contain a description of the time-dated schedule of practices or treatments from which, in turn, there can be derived the associated factor inputs. Next, as a general rule, since the forest manager generally comes from a biological management tradition, the outputs will be characterized through yield tables. These prescriptions and yields are the time-dated, hence dynamic, schedule of inputs and outputs, connected at least in part by time-dependent biological relationships within evolving plant communities. The set of such time schedules of inputs and outputs is precisely what the economist seeks to represent by the production function. What is required in order to convert the economist's convenient shorthand notation into practice is the forest manager's technical knowledge of the results of an alternative mix of treatments, or combination of activities, on the resulting stream of outputs and required basic inputs.

There are different approaches of more or less formality being taken to this problem, depending on practical circumstances. Generally the basic element is a silvicultural model that allows consideration of the effects of a range of prescribed treatments or practices for manipulating the forest vegetation. The resulting characteristics of the forest community, or habitat, are noted by drawing upon documented research results or other established knowledge if available, for example, stand simulation models, and informal conventional wisdom, or "professional judgment," where there are gaps in the documented knowledge, for each different combination of relevant plant communities. Other outputs may then be related to the changing forest condition. For example, a hydrologic simulator that relates water yield primarily by noting changes in evapotranspiration rates for plant communities of different composition may complement the silvicultural model. What of the implications of different compositions of forest communities for other attributes to the forest that may be of value to *in situ* uses and users of the forest? There is available, in the case of wildlife, for example, a combination of research findings and professional judgment useful for associating wildlife to habitat types, that is, for relating faunal relationships to plant communities of different characteristics (see, for example, Thomas, 1979; Verner and Boss, 1980). The detailed descriptions of variation in habitat types and of their association to faunal attributes of the forest community are comprehensive in one sense. However, the relationships described often tend to suggest only the effect a given change in vegetal cover will have on a habitat-desirability index for a given species, rather than on some measure of the level of the species population itself.

An example that attempts to incorporate all of the interdependent functional relationships into an inclusive quantitative simulation model is provided by Boyce (1977). Algorithms are developed that tend to provide the linkages or interdependent relationships between floral and faunal associations as a function of manipulating the dominant and codominant species. These then are integrated into a large-scale, relatively detailed computational model that intends to provide indices of the quantitative relationship between scheduled prescriptions and time profile of outcomes or the change in the level or mix of forest resource commodities and services, as a function of different combinations of activities, or schedules of either land treatments or forest practices or both out to an appropriate planning horizon.

Such work is significant in that it suggests that we may concentrate on a limited set of descriptive indices of the overall visual and habitat condition of forest areas and consider these, along with timber harvests, to be the forest outputs. We may accept the structural simplification with some confidence that the objects of actual consumer interest can be directly related to our indices of forest condition.

It is beyond our competence to judge how accurately the functional relation between prescriptions and the mix of outputs is described, or how adequately the present state of knowledge specific to this problem meets the aspirations of those responsible for passing the National Forest Management Act. The essential point is, however, that if one is required to consider the cost of factor inputs (and hence consider the budget and appropriations for managing the national forests) compared with the value of outputs, a relevant intervening mechanism that maps different combinations of activities into different mixes and levels of output is indispensible. Moreover, Forest Service scientists and management personnel are addressing this problem, and it is being addressed at a level of <u>conceptual sophistication</u> that is consistent with the complexity of the problem. A distinction needs to be made between <u>conceptual</u> sophistication and the refinement that is possible in actual application. Both gaps in scientific information and the exploding computational burden associated with increasing detail enforce practical limits in forest planning applications.

We may note that even with accurate simulation models of the forest dynamics available, any practical decision process involving land allocation will require the use of a simplified description of forest production potential. What we observe in practice is that a limited number of the many possible prescriptions for land use are evaluated through simulation. The linear programs that now aid in the final land allocation and scheduling decision may then choose from among a set of these candidate prescriptions. To consider a great many more such prescriptions would quickly tax the budget, staff, and computational facilities that can reasonably be allocated to planning. This raises the issue of how well such a limited sample of prescriptions can represent the known production possibilities of a forest. Clearly, the planning staff, by preselection of the candidate prescriptions, can strongly influence the quality of a final land allocation. Unfortunately, there does not seem to be any procedure in current use to ensure that the sample of prescriptions considered is representative of the potentially best land uses. Following is a rather informal look at procedures that may improve the set of candidate prescriptions.

The concern is that in determining the best combination of land uses to meet alternative forestwide output and service targets, the selected allocation is effectively constrained by the set of prescriptions entered into the linear program. It is almost inevitable that any given selected land allocation will be suboptimal[10] and could be improved upon by the introduction of some other candidate prescriptions. Ideally we would like to iteratively enter new prescriptions into the forestwide linear program so as to progressively improve our representation of production possibilities in the region of apparent optimality. We can do so by entering prescriptions that yield positive net benefits under the prices and shadow prices associated with the current optimal solution. At minimum we would like to screen out from consideration those practices that are unlikely to be appropriate under any reasonable set of output prices (and so, implicitly, under any reasonable constraints on flows of outputs).

As an example, for a few representative, small, geographically coherent units, such as lesser watersheds, it should be possible to consider a fairly complete range of prescriptions reflecting alternative roading schemes and various combinations of treatments across stands within the unit. By searching for a set of optimal prescriptions on these sample areas conditional upon various reasonable alternative demand-price assumptions, we may reveal a good deal about the nature of appropriate candidate prescriptions for other areas. The range of prices considered may reflect our relative uncertainty, with a wide range considered for the nonmarketed services and a lesser range for marketed goods. With the insight gained from such samples, an improved formulation of prescriptions for other similar forest areas should be possible, and we will have excluded from consideration any technically inefficient prescriptions. With this initial

---

[10] In the following discussion it should be understood that optimality and suboptimality are in reference to the given set of prices and constraints, and the particular formulation of the planning problem. This is in contrast to the more ambitious concept of social optimality.

screening completed, the forestwide linear program will be run using estimates of output prices reflecting current output levels to choose among the prescriptions for the subareas of the forest. The prices or shadow prices of the outputs in the forestwide solution may differ from those initially anticipated, because of constraints on budget and harvest flow or because forestwide output levels induce price changes. Based on these new prices, we will find it desirable to augment and improve the representation of the production possibilities by generating and including new prescriptions. We will be guided in this by the prices and shadow prices of the forestwide solution and may use the detailed sample area models to investigate the potential for improvement.

Such procedures cannot guarantee fully optimal land allocation. There does, however, seem to be room for improvement in the present procedure, and therefore it is suggested that greater attention needs to be paid to the wise choice of prescriptions selected to represent the production function.

It would perhaps be naive to suggest ready availability of all of the knowledge required for an accurate and adequate linkage of the transformation of inputs embedded in the prescriptions and time into the time profile of outputs. There will be need for additional, well-targeted research in certain areas. This is true particularly where rather small changes in the assumptions about what the actual (scientific) data on relations would look like if we knew them (that is, where they currently are inadequately understood) might result in a substantial change in either the mix, level, or timing of outputs. But the need for representation of the intervening mechanism relating the time profile of forest resource services to management inputs, as well as the practical approach taken in using this knowledge, suggests an important agenda for supply-side research relevant to national forest management.

THE CHALLENGE OF ACCURATE VALUATION OF FOREST RESOURCE COMMODITIES AND SERVICES

If our judgment regarding the direction and progress being made on the supply side is valid, then there is growing reason for optimism about the more technical (physical, biological, engineering--or supply) aspects of national forest management. Can the same thing be said about the demand side? Here, we believe, the breadth and depth of research on demand for the whole range of forest resource commodities and services by the Forest Service, probably related to its short history, in no way compares with its opposite number on the production side. And the deficiencies coupled with the inherent difficulties involved are very likely to be the fundamental challenge for efficient management of the national forests. Indeed, there has never been any question that foresters could determine the probable growth of trees and that wildlife biologists could identify the characteristics of habitats that tend to promote wildlife production and species diversity among forest and rangeland fauna. The remaining problem is one of determining the appropriate mix of current practices, given present expectations of the public's valuation of outputs and the opportunity cost of the inputs, including the land and budgetary resources that have alternative uses.

It is probably not surprising that the problem of adequately evaluating the demand for forest resource commodities and services should be the more difficult one. By reason of history, tradition, law, and policy, and because of the technical reasons that may preclude the supplier from excluding nonpaying consumers, various elements of the array of forest resource outputs fail to be distributed by means of pricing. This complicates the problem of efficient management, since the provision of nonpriced resource commodities and services from the national forests poses a real difficulty for preparing meaningful estimates of public benefits that are associated with forest plans, and for preparing the benefit side of the program required under the Renewable Resources Planning Act.

The role of prices in efficient management can be understood by considering how the land use decision might be made if adequate private competitive markets existed. If there were numerous sites to which access could be cheaply controlled and priced, the market would provide each owner with information on the prices that users should be charged for each service offered by the owners. These prices would depend upon the condition and accessibility of the owner's lands. In addition, a land market would develop with sale prices reflecting the anticipated streams of future net revenues both from harvest sales and from the marketing of other service outputs. Under such perfect market conditions, the land owner would consider the impact of any land treatment on both net revenues from harvests and services and the asset or sale value of his or her property. Market forces would then ensure that both present and future generations would receive as great a benefit from the supply of products and services as could be achieved by any public manager. In the absence of private competitive markets for certain service outputs, the public manager's problem is to estimate from observable economic behavior or other information the prices that should have been charged for access to the service outputs of each subarea of land--had there been in fact a competitive market clearing land allocation. Lacking the availability of direct market information, the manager must estimate these equilibrium prices through an analysis of the supply and demand equilibrium. Obtaining accurate and theoretically defensible estimates of these equilibrium prices for specific service outputs appears to be the major unresolved problem or challenge in formulating efficient plans for the management of the national forests. How might this problem be addressed in terms that are compatible with the flow of information stemming from analysis of the supply side in the forest planning process?

The impact of management is to alter the qualitative condition of a given forest area. The users of the output services will value such changes because of the incremental effects on the profitability or utility they may achieve as either producers or consumers. We concern ourselves here with techniques for valuing changes in the qualitative condition of recreational service outputs. There is a rather extensive literature both on the topic of the demand for recreational sites and on valuing qualitative changes.[11] One class of techniques uses information on the demands for related goods and services, such as travel. For example, multisite travel-cost models have been developed and used to estimate demands across a set of sites of varying quality. In the same vein, there is another class of models (hedonic models), which has not yet been extensively applied to recreation sites. Such models relate the demand for a private market com-

---

[11] A general review of approaches assessing benefits from improved environmental services is found in Freeman, 1979. For reviews of the methodology in recreation demand analysis, see Dwyer, Kelly, and Bowes, 1977; and also Smith, 1975.

posite good (like housing) to environmental quality characteristics such as air quality, which can be defined to be part of that composite good. A third type of approach is to elicit responses directly from users as to what they say certain levels of environmental, or, in our case, forest output services, are worth to them.

The following focuses on the first two classes of techniques, briefly discussing them in the context of valuing the forest output services used by recreationists. The first requirement is characterization of a discrete number of attributes or descriptors to represent different site conditions. These may be in terms of forest stand condition, diversity of habitat, accessibility, access controls, and perhaps special land form attributes that are significant to on-site users of national forest services. Then, in order to estimate the value of the different levels of the attributes that may result from alternative management practices, we must meet the following conditions: (1) It is necessary to observe a situation in which some consumers face different costs of using sites of given qualities. This condition is not very restrictive, because the different spatial allocation of users relative to the forest tends to guarantee its fulfillment. (2) There must be observable variability in attributes from one site to another as a result of past differences in management. (3) We must be able to observe consumers originating at different locations and choosing to use differing types of sites because of variations in the particular array of alternatives and in the cost facing each originating location (rather than choosing different sites because of taste differences). The existing methodologies most appropriate for estimating the magnitude of change in the benefits from one prescription to another are the multisite travel-cost model and the hedonic model. These are not inherently different from one another, and it can be shown that both can be derived from the same theoretical model of consumer behavior. They differ rather in the empirical assumptions that facilitate their application, and one might refer to the general approach as the hedonic/travel-cost model.

Multisite travel-cost models estimate demands for visits at sites that differ in such factors as the recreational service level, for example, due to different prescriptions applied (see Cesario and Knetsch, 1976). This approach values changes in the level of an attribute through evaluating shifts in the demand curve for site use. It is most appropriate when we can assume the demand for visits to sites would be identical except for the effects on the price and quality variables we have chosen to measure. We must then have observations of visitor use at sites with different attribute levels, and for each site we must observe use by visitors coming from various distances.

The hedonic approach, on the other hand, focuses directly on the demand for and value of an attribute.[12] Although it requires the same information as the multisite travel-cost approach, it fits best if there is a tendency for the number of visits to be relatively unaffected by attribute levels.[13] The demand curve for the attribute level is then estimated directly as a function of its marginal cost and the level of visits. The variation in the attribute level is valued through measuring changes in the area under its demand curve net of the cost to the consumer. Either of these two approaches can incorporate multiple attributes if they are necessary to describe the output services demanded by the recreationist.

While there are available methodologies upon which to build, a great deal of well-targeted research is required to bring the reservoir of documented knowledge in this area into balance with what is known through hydrologic, silvicultural, and bionomic research on the supply side of forest management. Even with an appropriately specific hedonic/travel-cost approach or a careful direct questioning approach there are special problems when habitat for game animals is under consideration. Several rather complex issues are involved, the successful approach to which will require a great deal of ingenuity, along with understandable trial and error. The case of hunting use presents some particularly challenging problems.

First, although skill, equipment, and choice of hunting site may affect the hunter's chance of success, bagging game is still basically a probabilistic event, where the variable the hunter takes into account in making decisions is "the expectation of bagging a deer." Individual hunter

---

[12] The theoretical basis for the hedonic approach is found in Rosen, 1974. Applications include Harrison and Rubinfeld, 1978; and Wilman, 1980.

[13] Wilman (1980) discusses this possibility in an application of the hedonic approach to assessing the value of beach quality. There the "length of visit" variable did not vary with the choice of beach quality.

observations, however, only represent outcomes, and must be averaged together to reconstruct the hunter's prior expectation.

Second, under our federal system the management of game populations (unless they are migratory) is left with state agencies. Accordingly, a combination of management practices is required--namely, habitat management by the Forest Service and game management and access regulation by the relevant departments of fish and game at the state level. The estimation method must take into account the effect these game management regulations have on hunter behavior as it is described in the benefit estimation model. If, for example, hunters are constrained either to a given season length or to only one deer or to both, the observed number of visits may be constrained by those limitations--as may the level of harvest--and the effect this has on choices must be incorporated into the theoretical model that will be estimated using observations on hunter behavior. Admittedly this is a special case because two different agencies at different levels of government need to be involved in coordinated management of biological resources that give rise to recreational activity. Many other recreational activities are directly related to access, characteristics of the recreational site or area, or other features resulting from land management over which the Forest Service has exclusive jurisdiction, making the problem less complex than indicated above.

An additional problem does not directly affect the estimation of equilibrium prices, but has implications for the efficiency of resource management if use is not rationed at a level consistent with these prices. Let us assume that every characteristic of the forest over which the forest land manager has control will give rise to a flow of services, the benefits of which, whether on- or off-site (timber or minerals as off-site resource commodities) can be accurately measured in monetary terms. Then, by playing out various sets of feasible prescriptions, the large-scale linear programming model used by the Forest Service to identify the particular combination of activities or prescriptions that has the highest present net benefit can be found. We will then have, in the language of the operations research fraternity, the "optimal forest plan." However, if all of the services provided by the optimal forest plan, while capable of measurement, are not in fact subjected to properly levied user charges, then an inconsistency between the services capable of being provided in a cost-effective manner and the demand for the services (at zero price) will develop. The resource commodities or services for which charges are not being made will be "oversubscribed," and if they are services that people go to the site to enjoy, there will be excess demand--in other words, there will not be enough capacity to service the demand--even though the additional capacity could not be provided at costs justified by the marginal users' valuation of the benefits. There would be a fundamental inconsistency between the economically efficient forest plan and the public demand at zero price with quality deterioration (congestion), perhaps resource degradation, and a wholly unnecessary management problem. Anyone familiar with some Forest Service recreational areas will have no difficulty recognizing the problem. The plan may be efficient, but the management (policy regarding user charges) may be inconsistent with such a plan to produce the benefits attributed to it. Given the tradition in public lands management, it will not be easy to come to terms with this very troublesome problem (although it is in fact a simply resolved pricing problem often confused with the ecological concept of carrying capacity).

MANAGEMENT AND BUDGET ISSUES

While valuation of the forest resource services, especially over time, is a challenging problem, we are convinced that it will yield as readily to research and study as the problems associated with forest hydrology, silvics, and bionomics. Moreover, well-designed, carefully targeted research will permit the investigation of economically efficient forest plans at different plausible price (value) schedules within the realistic range to check for what differences would be indicated in the "optimal forest plan" with parametric adjustment. In short, it is conceivable that the immediate prescriptions for a given forest might be relatively insensitive to changes in relative prices within the relevant range, and to learn this would be of value.[14] Apart from introducing variation in price schedules for sensi-

---

[14] Of course, only the immediate prescriptions are of real interest. As information is revealed over time, it will be found desirable periodically, as also required by law, to revise the original plan through a new set of "immediate prescriptions."

tivity analysis, the computational models can be used to address the question of the desirable level, and potential change in mix of forest outputs, for different plausible budget or appropriation levels. It must be recognized that a different appropriation of cooperating (budget) resources which provide for the capital and current expenses needed to achieve any given level of production may result in different relative proportions among the resource services covered by the Multiple Use Act; it is a virtue of the programming models that they can be used to investigate this aspect of resource allocation as well as the allocation of land to different purposes.

Before addressing this problem directly, we need to consider a preliminary issue. Let us assume that we can obtain reasonably good approximation of the time profile of the changing mix of outputs (yield tables) as a result of a given prescription, get the present value of the factor costs (economic tables), and estimate of benefits. Precisely how will this be used? One important realization is that the output from a given prescription is a joint product in a very fundamental sense. There may be some activities, undertaken in any prescription, that can be accounted for exclusively by the inclusion or exclusion of an increment of a separate forest resource product or service, but many activities will simultaneously serve two or more purposes. There is no nonarbitrary way of separating the portion of the costs attributable separately to each. Does this pose any problem for land use planning? The answer is no--as perhaps was first established by the late Professor Wantrup more than a quarter-century ago in addressing this problem in the water resources field. If we had the present value of costs for each prescription or program, along with its associated present value of benefits, then all that would be required would be to compare the present costs and benefits of each of the numerous different prescriptions and select the prescription for the forest that would maximize the positive difference between benefits and costs. (At the moment we abstract from problems associated with budget or capital constraints in the selection of the optimal program.) Since these prescribed activities apply to specific parcels of land, we have in each prescription a different implicit allocation of land to different purposes to obtain different mixes of outputs. The land allocation is an implicit part of the solution by the appropriate selection of the benefit-maximizing prescription. Indeed, the incomparably complex problem of common cost allocation, for which there is no theroretically valid method where joint products are involved, is not an issue in planning the optimal use of a national forest. To the extent that Forest Service personnel are embroiled in such a problem, it will have been for some reason extraneous to the proper planning of forest land uses.

Now it is true that in order to manage access to the resource commodities and services of the national forests, a configuration of relative prices needs to be established. That configuration will limit the actual quantity demanded to the amount of each commodity or service that can be efficiently supplied according to the optimal plan in the planning process. Of course, some method other than price could be used as a rationing device if a compelling case could be made for it, but it is not clear that alternatives to pricing or charging user fees for all forest resource commodities or services are equally efficient. The problem is not one of attempting to allocate common costs, but simply one of making certain that the value of any increment in a given resource output covers its incremental attributable costs and that, over all, the benefits remaining exceed the common costs of the aggregate. That, of course, is achieved automatically with the choice of the forest prescription (plan) that has total benefits from a particular mix of commodities and services exceeding total costs by the largest margin. This assumes, of course, that the prescriptions from among which one is selected have all been cost effective or technically efficient.[15]

---

[15]Perhaps it is useful to distinguish between "economically efficient" and "cost effective." Cost effective means that the least-cost means and combination of factor services have been selected to achieve any given purpose. It does not mean that the purpose itself represents an economically efficient allocation of resources. That is, the benefits from the cost-effective prescription may not justify the cost. However, if the least-cost method to achieve the objectives of a given prescription has been employed, and from among such cost-effective prescriptions one enjoys a greater positive difference between total benefits and total costs, that prescription is not only cost effective, it represents an economically efficient allocation of resources, that is, it is both cost effective and economically efficient.

Accordingly, complex cost allocation addressing common costs is not a problem. The problem is rather to establish a system of user charges, fees, stumpage prices, and so forth, that will clear the quantities being supplied by the previously determined optimal forest plan. This is simply stating in another way the problem raised earlier with reference to the need for user charges in order to eliminate the inconsistency between the supply of resource commodities or services that can be provided in an economically efficient manner, and the excess demands that can be guaranteed when the outputs are subsidized or provided free of charge. This is a fundamental challenge for efficient management of the national forests.

Another serious management problem relates to the way in which the accounting, budgeting, and appropriations process is presently conducted. We are accustomed to visualizing forest management programs in terms of the several purposes specified in the Multiple Use Sustained Yield Act of 1960. Indeed, the programs for National Forest System management are characterized in these terms (namely, range, recreation, timber, water, wilderness, and wildlife), and the program elements for budgeting largely follow these lines. However, as mentioned before, the prescribed activities, or forest management programs, involve joint production with a large measure of common costs. They are not meaningfully separable. But for budgetary and appropriations purposes a format has been adopted that requires an arbitrary allocation of the budget among the conventional line items. If personnel at the Office of Management and Budget (OMB), Congressional Budget Office (CBO), and the appropriations and related committees did not impute some direct relationship between the cost allocation among the line items and the output from a given line item program, no harm would be done. But the tendency has been for the appropriations to be made independently by line item; it is surprising then if the requested bundle of line items can be efficiently provided by the resulting appropriation. An appropriation might provide more than is needed for efficient provision, or, as is more likely, require provision of a bundle of line items that cannot be produced with that budget because it violates the biological relationships inherent in the production of the array of forest resource commodities and services.

Accordingly, there is need to recognize the problem and to develop a more realistic budget format that would be both consistent with the realities of biological processes in forest output production and conducive to better organization of budgetary information for management purposes. Such a system should reflect at least two important elements:

1. A sufficiently diverse mix of alternative prescriptions (alternative forest plans) for any given budget level on a given forest so as to provide a choice from among several different mixes of prescribed practices and their associated output mixes, and

2. A similar set of diverse alternative forest plans for different realistically plausible budget levels.

Such an array of alternatives, when aggregated regionally and nationally, would permit choices among alternatives that would relate different outputs associated with different prescriptions and budget resources (appropriations). Personnel at OMB, CBO, and the relevant committees of Congress would then have the information that would allow funding (for whatever level and mix of outputs from the National Forest System) to be consistent with the dynamic interdependencies in biological/economic production relations in the national forests. Under the existing set of circumstances, Congress can, by an appropriations act, unwittingly nullify the mandate it gave the Forest Service under the National Forest Management Act.

CONCLUSIONS

It is our view that the process initiated by the Renewable Resources Planning Act and the National Forest Management Act pursuant to multiple uses of forest land and rangeland is progressing, not only in the right direction, but also with admirable accomplishment, given the monumental size of the task in relation to the time and budget constraints. The level of intense effort necessary to launch a systemwide forest planning exercise is doubtless the most demanding the first time around, but the achievements speak well for the effort.

There are numerous areas that we would like to see improved before a second round of updated plans is undertaken. One of the problems is to provide a wide enough array of either alternative prescriptions or plans or both, so that meaningful choices can be made. We have suggested that a more detailed review of diverse prescriptions on representative areas might be undertaken as a "preprocessor" activity to enable the computer program to select the range of optimal prescriptions for each of several realistically plausible budget levels.

A second area in which we would like to see improvement has to do with necessary research directed at the scientific information essential to building the linkages between the scheduled activities or vegetation manipulations and the time stream of resulting outputs from these activities. This is a precondition for an accurate accounting for inputs and outputs and in the treatment of the costs and gains associated with management activities on the supply side.

Thirdly, we believe that innovative research directed toward improving the quality of the demand estimates in realistic economic terms is among the highest priority areas for additional work. Our guess is that the realism of the forest plans is most likely to be suspect in valuation of nonpriced resource services. Some conscientious attention by both the Congress and the executive branch policy makers to the question of compatibility between the planned outputs and the terms on which access to such resource services is obtained--namely, user fees--is very critical to both improved planning and efficient management.

Finally, the entire issue of allocating common costs needs to be recognized as a false issue and as one that is likely to create a great deal of mischief unless it is seen in proper perspective and unless a sensible means to deal with it is devised. Here it is important to provide a budget/appropriations format that will accurately reflect the implications of one or another budget or appropriations level, rather than to have arbitrary accounting procedures disguise the true effect of a given budgetary or appropriations response to Forest Service program requests.

REFERENCES

Boyce, Stephen G. 1977. *Management of Eastern Hardwood Forests for Multiple Benefits* (DYNAST-MB) Forest Service Research Paper SE-168 (Asheville, N.C., Southeast Experiment Station).

Cesario, F. J., and J. L. Knetsch. 1976. "A Recreation Site Demand and Benefit Estimation Model," *Regional Studies* (10 March), pp. 97-104.

Chan, M. W. Luke. 1981. "A Markovian Approach to the Study of the Canadian Cattle Industry," *Review of Economics and Statistics* vol. 63, pp. 107-116.

Clawson, Marion. 1978. "The Concept of Multiple Use," *Environmental Law* vol. 8, no. 2, pp. 282-286.

Dupuit, Jules. 1844. "De la Mesure d'Utilite des Travaux Publique," in *Annales des Ponts et Chaussees* (Reprinted under the heading "De l'Utilite et de sa Mesure," in *La Riforma* (Turin, 1932)).

Dwyer, J. R., J. R. Kelly, and M. D. Bowes. 1977. *Improved Procedures for Valuation of the Contribution of Recreation to National Economic Development* (Urbana-Champaign, University of Illinois).

Fisher, I. 1954. *The Theory of Interest* (New York, A. M. Kelley).

Freeman, III, A. Myrick. 1979. *The Benefits of Environmental Improvement* (Baltimore: Johns Hopkins University Press for Resources for the Future).

Haigh, John A., and John V. Krutilla. 1980. "Clarifying Policy Directives: The Case of National Forest Management," *Policy Analysis* (Fall).

Harrison, Jr., David, and Daniel J. Rubinfeld. 1978. "Hedonic Prices and the Demand for Clean Air," *Journal of Environmental Economics and Management* vol. 5, pp. 81-102.

Hirshleifer, J. 1970. *Investment, Interest and Capital* (Englewood Cliffs, N.J., Prentice-Hall).

Jarvis, Lovell S. 1974. "Cattle as Capital Goods and Ranchers as Portfolio Managers: An Application to the Argentine Cattle Sector," *Journal of Political Economy* vol. 82, pp. 489-520.

Krutilla, John V., and Otto Eckstein. 1958. *Multiple Purpose River Development: Studies in Applied Economic Analysis* (Baltimore, Johns Hopkins University Press for Resources for the Future).

Krutilla, John V., and John A. Haigh. 1978. "An Integrated Approach to National Forest Management," *Environmental Law* vol. 8, no. 2.

Malinvaud, E. 1953. "Capital Accumulation and Efficient Allocation of Resources," *Econometrica* vol. 21 (April), pp. 233-268.

Musgrave, Richard A. 1959. *The Theory of Public Finance: A Study in Public Economy* (New York, McGraw-Hill).

"Report from the Secretary of the Treasury, No. 250, to the First Session of the 10th Congress, April 6, 1808." 1834. *In* W. Lowrie and W. Franklin, eds., *American State Papers: Class X, Miscellaneous* vol. I (Washington, D.C., Gales and Seaton), pp. 724-921.

Rosen, Sherwin. 1974. "Hedonic Prices and Implicit Markets: Product Differentiation in Pure Price Competition," *Journal of Political Economy* vol. 82, pp. 34-50.

Shands, William, and Robert Healy. 1977. *The Lands Nobody Wanted* (Washington, D.C., Conservation Foundation).

Smith, Adam. 1976. *An Inquiry into the Nature and Causes of the Wealth of Nations*, edited by Campbell and Skinner (Oxford, Clarendon Press).

Smith, V. Kerry. 1975. "The Estimation and Use of Models of the Demand for Outdoor Recreation," in *Assessing the Demand for Outdoor Recreation* (Washington, D.C., National Academy of Sciences).

Spreen, Thomas H., Bruce A. McCarl, and T. Kelly White. 1980. "Modelling the Cattle Subsector: A Demonstration in Guyana," *Canadian Journal of Agricultural Economics* vol. 28, no. 1.

Thomas, Jack Ward. 1979. *Wildlife Habitats in Managed Forests: The Blue Mountains of Oregon and Washington*. Agricultural Handbook No. 553 (Washington, D.C., USDA, Forest Service).

Verner, Jared, and Allen S. Boss. 1980. *California Wildlife and Their Habitats: Western Sierra Nevada* (Berkeley, USDA, Forest Service, Pacific Southwest Station).

Wilman, Elizabeth A. 1980. "Hedonic Prices and Beach Recreational Values," in V. Kerry Smith, ed., *Advances in Applied Micro Economics* (Greenwich, Conn., JAI Press).

Clark S. Binkley

COMMENTS

The long production process associated with forested ecosystems forces those concerned with managing their multiple outputs to plan. Certainly what Hayes (1969) has called the rational paradigm for resource management extends deep into the intellectual roots of the forestry profession, perhaps culminating historically in the idea of normal forests arranged to yield the maximum physical quantity of wood in perpetuity. The rational paradigm has found happy pasture in recent federal legislation governing planning on the national forests and national planning for renewable resources in general.

In chapter 6, Krutilla, Bowes, and Wilman review economic planning for the national forests, together with its underlying institutional and political infrastructure. They conclude that the National Forest Management Act (NFMA) requires the Forest Service to attend to economic efficiency in national forest management. Further, land use planning is the mechanism for achieving this desideratum. In their view the Forest Service is addressing the problem "at a level of conceptual sophistication that is consistent with the complexity of the problem." They conclude by emphasizing research needs on the supply but more importantly on the demand side, and by attempting to lay to rest once and for all the joint cost problem.

We should apply the rational paradigm not only to national forest plans, but also to the plans for planning. There are at least two dimensions to the strategy for planning: one relates to the choice of analytical technique and the other to the optimal application of the appropriate techniques. In the first case, should we use linear programming to model a decidely nonlinear system? In the second and perhaps more important case, do the benefits of the current planning system exceed the costs?

IS LINEAR PROGRAMMING APPROPRIATE?

Forested ecosystems are notoriously nonlinear systems, and the optimal management of them stretches the assumptions of linear programming embodied in FORPLAN. The detail to support this assertion would fill at least a forestry curriculum, and perhaps several lifetimes of research and experience. I hope one example will suffice to give the flavor of my concern. Consider perhaps the most simple multiple use problem imaginable--the optimal rotation of a loblolly pine plantation where the only products of concern are pulpwood and sawlogs. It is easy to show that in general two "optimal" rotations exist, roughly one corresponding to the peak value of the stand for pulpwood production, and the other to the peak value of the stand for sawtimber production. Which of these local present net worth maxima is the global optimum depends on the discount rate and the relative prices of sawtimber and pulpwood. Existing harvest scheduling algorithms simply assume the existence of a single turning point, so that any local maximum that is found is assumed to be the global solution. Indeed, the iterative procedure outlined by Krutilla and his colleagues would probably converge on one of the local optima in the multiple use problem, but we have no guarantee this solution would be the best management plan.

THE COSTS OF PLANNING NFMA PLANNING

Simon (1978) argues that the objective of procedural rationality is to find the least-cost or best-return solution decisions _net_ of the costs of information, computation, and other analytical effort. Recently Row (1981) summarized the enormous analytical cost of the forest level plans, including training hundreds of people in the use of linear programming, revising and extending resource inventory information, establishing and operating the interdisciplinary teams on each national forest to formulate the land management problem in a linear programming setting, and bearing the computational burden of actually solving the problem once it has been formulated and the requisite "data" have been assembled and inputted.

Consider the recent plan for the Lolo National Forest in Montana. Informal estimates of the plan's direct costs range around $1 million. That plan involved forty-eight Forest Service employees or contractors, and perhaps 1,000 public participants (USDA, 1980). Because personnel ceilings constrain total Forest Service employment, the cost of this labor might exceed the sum of wages paid. No doubt the costs of NFMA planning will decline as more experience in the process is developed, but the costs should be monitored and periodically weighed against the benefits.

While the conceptual economic underpinnings for the multiple use problem are straightforward (Bowes, 1980), the scientific understanding of the production trade-offs is weak. Although the general nature of resource interactions seems reasonably well understood, a substantial research effort is required to provide the degree of quantification necessary for the application of linear programming to each of the national forests in the country.

In addition to the direct costs of analysis, there are effects on the institutional structure of the Forest Service. Fairfax has argued that these might be significant, taking land management decisions from land managers and placing them with "lawyers, computers, economists and politically active interest groups . . ." (Fairfax, 1980, p. 206). While it might seem improvident for an economist to condemn activities that increase the demand for economists, and presumably their wage rate, I share this concern with Fairfax. Less effective management will eventually translate these institutional effects into economic costs.

THE BENEFITS OF PLANNING

As is frequently true in problems of resource economics, the costs dawn more clearly than do the benefits. Broadly speaking, the benefits of planning are political and functional (abstracting from the fact that for the Forest Service, politics may be function). Concerning politics, the planning process establishes a formal procedure for reconciling conflicting interests in the various possible multiple outputs of a specific national forest. Thus the distributional implications of alternative policies are displayed and incorporated into the process of evaluating land management alternatives. Marion Clawson has repeatedly pointed out the possibility of simultaneously increasing both extractive and <u>in situ</u> uses of the national forests through more effective management. The planning process offers an institutional structure for negotiating such Pareto improvements. Any such movement toward productive efficiency should be applauded even if the result is not economically optimal.

No doubt the political benefits of the planning process transcend this essentially economic function. Leman (1980) suggests that these benefits include an improved stance for negotiating budgets in Congress and insulation from competing interest groups. But while the political benefits might be great, let us focus on the functional ones.

The value of any information activity, planning included, is the avoidance of the costs associated with incorrect decisions. In a functional analysis, incorrect decisions occur because as near as we can measure or forecast, the world remains uncertain. In this context, planning serves decisions by recognizing how current actions affect the opportunities for future production from a resource input. Were outputs not linked through time, planning would lose most of its economic necessity: why worry about cutting a tree this year or next if a tree could be produced in one year's time? Further, were the world deterministic (or more precisely, were we able to anticipate perfectly the future conditional on some specific action taken), these intertemporal linkages would not be particularly troublesome. In that case time would simply be an input that should be applied until the value of its marginal product equaled its marginal cost.

The linkages through time become critical in the realization that the future of the world is truly unknown, and that the capacity to foresee the future in response to a current action is quite limited. Thus a purportedly optimal management plan is optimal only with respect to some specified--and it would be hoped likely--future state of the world. In fact, the "optimal" management plan might be quite bad under some likely combinations of production trade-offs and output values. In short, our deterministic planning process tends to reject policies that are good with respect to many possible futures but optimal in none in favor of policies that are optimal in some future, but whose performance in light of other futures is unknown.

The concept of economic regret is useful in this situation. That is, how well does a policy perform under circumstances other than those for which it was designed? For example, suppose two future states of the world, denoted I and II, are considered possible. These might correspond to high and low values for nontimber outputs. The planning process is considering three alternatives, unimaginatively named A, B, and C. The present net worth of the forest outputs under each plan can be computed for each state of the world, and suppose these are as given in table C-6-1 under payoffs.

Table C-6-1. Present Net Worth and Losses of Forest Output Under Three Alternative Plans for Two Possible Futures

State of the world	Payoffs for			Regrets for		
	Plan A	Plan B	Plan C	Plan A	Plan B	Plan C
I	1,000	400	900	0	-600	-100
II	200	800	700	-600	0	-100

Regret is defined as the loss associated with an incorrect decision. In the table, for state of the world I the optimal plan (the decision with highest present net worth) is A, and for state of the world II the optimal plan is B. If policy A is selected and state II occurs, opportunity loss of 600 occurs. This is found by subtracting the payoff from policy A in state II (200) from the payoff under the optimal plan (that is, plan B) in that future (800). This is the loss associated with incorrectly assuming that state II would occur. Of course, the regret from state I under plan A is nil because that policy is optimal for that state.

The notion of regret is useful both in defining the criteria for selecting a plan, and in measuring the benefits of the planning process. In the first case, policies with low regret are preferred to those with high regret, all else being equal. Note that policy C in the table is optimal in neither state of the world, but has low regret in both (in fact, C is a "minimax" policy--it minimizes the maximum loss). For this reason, plan C has some claim on the term "optimal," but policies of this type would not necessarily be identified under the current NFMA planning procedures. The presence of multiple local optima would exacerbate this problem.

In a narrow sense, the benefit of the NFMA planning process can be measured by the regret of the plan existing prior to the NFMA. That is, what is the difference between the present net worths of the existing forest plan and the plan selected through the new planning process? The data presented in the Lolo environmental impact statement are not adequate to make the calculation for that specific instance, but the output vector from the selected plan seems to differ little from that of the selected alternative. Consequently this measure of benefit would probably reveal little value in the NFMA planning process.

Recognizing the uncertainty in forecasts underlying the "optimality" of any plan leads to another concern. The deterministic planning process ignores adaptive responses to unforeseen events. As these events are revealed, policies will logically shift in response to the new situation and to altered perceptions of the future. Realizing that the desirable course of action may shift over time confers value on current decisions that leave room for adaptive responses in the future. Indeed, deterministic and adaptive decision models can lead to quite different first-period decisions (Grossmann, 1977).

Perhaps this would all be clearer with a few examples of what are probably adaptive and maladaptive actions. Irreversible decisions are the best general example of maladaptive policies. With timber production, cutting a tree is slowly reversible, so price uncertainty conceivably lengthens the optimal economic rotation age (Norstrom, 1975). Thinning is probably an adaptive policy. Judicious manipulation of the growing stock can keep average stand diameter high while sacrificing only modest cubic volume production. Such a stand would have value either as sawtimber or as biomass, where an unthinned stand might be suitable only for biomass. That the list of my examples is no longer and its quality no better testifies to the feeble development of this notion of robust policies in forest planning.

An adaptive plan is a consensus of actions to be taken conditional on specified alternative futures. Krutilla, Bowes, and Wilman go part way down this road, recognizing that different "optimal" plans will result from alternate budget constraints and resource values. New information on resource interactions might affect the optimal plan as well.

An adaptive planning process would monitor deviations from forecasts and would respond expeditiously to those changes. While monitoring seems to be an integral part of NFMA planning, the ability to respond to new information may be limited by the institutional structure of the public participation process. Specifically, public participation focuses on developing consensus around a single course of action. Presumably when circumstances change enough to warrant a new plan, the whole cumbersome process will be repeated. This would seem to doom the Forest Service to indecision. A cynic might see that as one of the political objectives of planning.

CONCLUSION

National forest plans consider economic trade-offs, but the planning process itself seems to ignore them. The costs--economic and institutional--of the process seem high, and the data on production trade-offs seem too limited to support the formal, quantitative planning models in use. Given adequate scientific information and computational capacity, this approach might be tenable, but it would still fail to focus on the important time linkages that are critical in an uncertain world.

But decisions must be taken in this complex and stochastic world. As an alternative, consider using the adaptive properties of markets themselves as a planning mechanism. Because I am skeptical about our current scientific ability to model in a quantitative way the production trade-offs inherent in multiple use management, I believe that that part of the analysis should be left to the education, experience, and wisdom residing in the computer lying between the ears of each forest supervisor or district ranger. To achieve the required reckoning of economic trade-offs, let each forest be made a profit center in the accounting sense. And let the national office set "transfer prices" for extramarket resources such as recreation and sediment loads at something approximating their marginal social value. Assessment should be made of realistic costs of the inputs used by each forest--land, timber inventory, costs of labor and equipment, and the costs of running FORPLAN. The multiple outputs of timber, recreation of several types, water quality, and so forth, should be monitored, and the forest supervisor's performance evaluated on the basis of the net return.

With this approach, land management decisions would rest squarely with land managers. To be sure, much of the same information and analysis currently a part of the NFMA planning process might persist, but its application would focus on substance rather than procedure. Plans would assume their fundamental roles of positioning management to deal effectively with uncertainty. Efficient allocation of resources, to forest production and to planning, would result.

REFERENCES

Bowes, M. D. 1980. "Multiple Use Planning with Non-commodity Services." Paper presented at International Union of Forest Research Organizations/ Man and Biosphere conference on Multiple Use of Forest Resources, May 18-23.

Fairfax, Sally K. 1980. "RPA and the Forest Service," in W. E. Shands, ed., A Citizen's Guide to the Resources Planning Act and Forest Service Planning (Washington, D.C., Conservation Foundation).

Grossmann, D. C. 1977. "Capacity Expansion for Water and Wastewater Services." Unpublished Ph.D. thesis, Massachusetts Institute of Technology.

Hayes, S. P. 1969. Conservation and the Gospel of Efficiency. (New York, Atheneum).

Leman, C. 1980. "Resource Assessment and Program Development: An Evaluation of Forest Service Experience Under the Resources Planning Act, with Lessons for Other National Resources Agencies." Unpublished. Office of Policy Analysis, U.S. Department of the Interior, Washington, D.C.

Norstrom, C. J. 1975. "A Stochastic Model for the Growth Period Decision in a Forest," Swedish Journal of Economics Vol. 77, pp. 329-337.

Simon, H. A. 1978. "On How To Decide What To Do," Bell Journal of Economics vol. 9, pp. 494-507.

Row, Clark, 1981. "Application of Systems Analysis in U.S. Forest Service Assessments and Planning" in P. Uronen, ed., Proceedings of the Forest Industry Workshop, IIASA Collaborative Paper CP-81-3.

USDA. 1980. "The Lolo National Forest Plan Draft Environmental Impact Statement." Forest Service, 01-16-80-01 (Missoula, Mont.).

Chapter 7

IMPACTS OF THE RPA/NFMA PLANNING PROCESS
ON MANAGEMENT AND PLANNING IN THE FOREST SERVICE

Douglas R. Leisz

It has been nearly seven years, as of this writing, since President Ford signed the Forest and Rangeland Renewable Resources Planning Act (RPA) into law, and little more than four since the National Forest Management Act (NFMA) amended it. During those years the Forest Service has made considerable progress in implementing the comprehensive planning process that these laws established. Early in 1980 the agency sent to the Congress the products of the second cycle of RPA planning: the 1980 Assessment (USDA, Forest Service, 1980b) and Program (USDA, Forest Service, 1980a). In September 1979, the Forest Service published land management planning regulations under section 6 of the National Forest Management Act, and is well under way with the first round of regional and forest planning under those regulations.

The impacts of this process on Forest Service management and planning are not all easy to identify. Since the planning process is still being implemented, even some of the immediate impacts are not obvious, and other impacts will develop as the Forest Service adapts organizationally to full implementation of the planning process.

RPA and NFMA became law as part of a general rechartering of the federal resource management role. When trying to pick up any thread of that role, one finds it part of an intricate tapestry of federal resource policy, so that it is difficult to set out a clear, causal relationship between any one legislative provision and a subsequent change in agency behavior.

Changes in law prompt other changes as well, and significant change requires tremendous amounts of organizational energy as agencies adapt to new directions. However, the full extent of change cannot be accurately assessed until the case law is developed and the laws are fully implemented. In the case of the amended RPA, the true impacts will not be known until Forest Service field personnel identify the practical aspects of carrying it out. Therefore it is too early to tell if all of our great expectations of this RPA/NFMA planning process will be fulfilled in a practical, operating sense. Some even suggest that it simply cannot be done.

This chapter is a preliminary assessment, which first reviews some of the significant changes prompted by the RPA and NFMA and the difference they have made in Forest Service management and planning, and then suggests some further changes in that planning process to make it less complex, less difficult, and less costly to carry out.

SIGNIFICANT CHANGES RESULTING FROM RPA/NFMA PLANNING PROCESS

There is little question that Congress has brought about significant national change through the Renewable Resources Planning Act of 1974 and the National Forest Management Act of 1976. It has occurred not only in Forest Service management and planning, but in other areas as well--in the Department of Agriculture, in other departments and executive branch agencies, and in the Congress.

Renewable Resources Planning

Probably the most far-reaching aspect of this change is that the RPA and NFMA have led to strategic and long-term planning for renewable resources on a national scale. For the first time in its recent history the Forest Service has attempted to take stock of most of its renewable resources and to decide how it wants to manage and use them for the future. The Forest Service's RPA Assessment, prepared with no particular agency program in mind, provides an objective look at the renewable resource situation nationwide. (With a few adaptations, this Assessment could form the basis for renewable resource planning in any natural resource agency or organization.)

As the first multiresource view of forest and rangeland conditions and trends, the Assessment has become a basic source of data and information for all people and organizations interested in natural resources. The RPA Assessment has focused national discussion of resource issues on a few generally agreed upon trends and on the needs and opportunities that are displayed in it.

This extensive use of the RPA Assessment has caused the Forest Service to reassess the adequacy of its inventory processes, the comparability of the several data bases used, and the utility of the information they yield.

This taking stock of the nation's renewable resources has shown that many of the decisions made intuitively by the Forest Service in the past were good, sound decisions; it has helped the agency better identify how one resource interacts with others, particularly in areas where the interactions had not been evaluated; and it has helped identify gaps in knowledge and technology that research should address. As a result of such information, decisions on the ground can be improved.

RPA progeny includes the Soil and Water Resources Conservation Act (RCA), the Renewable Resources Research Act (RRRA), the Renewable Resources Extension Act (RREA), the Cooperative Forestry Assistance Act (CFAA), and the presidential directive on program development for the Bureau of Land Management--which together nearly comprise governmentwide resource planning at the federal level.

The planning process has also encouraged state-level resource planning. California and Maryland have enacted planning legislation similar to the RPA, and many other states have expanded their resource planning under existing authorities and with federal financial assistance.

The RPA/NFMA mandate has encouraged coordination of federal and state resource planning. Land and resource management planning regulations promulgated under section 6 of the NFMA recognize that the most effective management and use of renewable resources occur when policy development, planning, and administration are coordinated among the several levels of government and with the private sector (Cooper, Wise, and Shands, 1980).

Central Focus for Policy and Programs

A second major change due to the RPA/NFMA process is a new central focus to policy and program direction. Prior to 1974, land management planning was done locally under various regional guides and was not tied to a comprehensive national planning, programming, and budgeting process. Now, however, programs are formed with national objectives, using locally derived data.

This national perspective based on locally assembled data provides a picture of resource demands and opportunities that was not possible when planning was only done locally. This new bottom-up, top-down approach requires forging some additional links. As a result, programs are built nationally according to national opportunities and needs, tempered by budget realities, and applied locally according to local capabilities.

The RPA/NFMA process provides an umbrella for Forest Service planning and management systems. Everything else--budgets, programs, and so forth-- is related to the RPA planning process. The best indicator of this is that, in preparing and submitting the RPA Program, the Forest Service faces much the same scrutiny as it faces in the budgeting process (Hagenstein, 1980). The RPA Program has received so much attention from the Office of Management and Budget because it is considered the driving mechanism for the Forest Service budget request and an influential guide to Forest Service programs for the congressional appropriations process.

This leverage on the budget can be quite troublesome during times of fiscal austerity. It promotes heartburn among budget builders charged with keeping spending down, because it displays the nation's future needs, along with supply opportunities. It can make a solid case for expanded investments in natural resource management, but can also serve as the basis for budget reductions, since trade-offs can be measured over the longer term.

Enhanced Analytical and Interdisciplinary Skills

A third area of change due to the RPA/NFMA planning process is an accelerated trend toward a broader mix of professional skills and greater analytic capacity within the Forest Service.

From January 1972 to January 1980, the number of foresters and engineers employed by the Forest Service declined by 8 percent. In the same time period there was a substantial increase in the number of economists, operations research analysts, computer scientists, soil scientists, biologists, and landscape architects. While this is not entirely due to the requirements of RPA/NFMA planning, it is an indication of an improved interdisciplinary mix of professional skills and an addition to the agency's general analytic capacity.

Furthermore, the broader requirements of the process have led to reorientation of Forest Service research, lending greater emphasis to multiresource and integrated management concerns. RPA brought about a thorough analysis of forestry research efforts and their relevance to the selected program direction.

This broadening of the Forest Service's professional disciplines will continue to shape the agency and its decision making for decades to come. It could mean broader, better-integrated, and more analytical management decisions for planning day-to-day operations and monitoring and evaluting agency work.

WHAT DIFFERENCE DOES RPA/NFMA PLANNING MAKE?

The changes discussed above are leading to some significant differences in Forest Service management and planning. The differences will be more evolutionary than revolutionary, becoming more apparent over the longer term. Any expectations of rapid or radical changes in policy or programs must give way to the more practical realities of making incremental adjustments in programs as they continue to operate at full tilt. The following sections describe several differences that stand out already.

Improved Analytical Approach to Planning

The first major difference that RPA/NFMA planning has brought about is that managers can now deal with a much more complex world. Analytical tools have enabled them to recall and consider more factors, to consider a broader range of options, and to display the anticipated consequences of each over time.

The Renewable Resources Planning Act and the National Forest Management Act provide a planning structure that is decidely economic (Haigh and Krutilla, 1980). It includes an assessment of demand and supply, a comparison of costs with benefits, and a requirement that economic analysis be used at all appropriate places throughout the planning process. Thus, the land and resource management planning regulations contain frequent references to economic analysis. In my view, the result of all this is a planning process for a public resource agency that leads to rational, economically sound analysis tempered by professional judgment in decision making.

The RPA, the NFMA, and the land and resource management planning regulations also provide for a wide range of social/political considerations.

They bear liberal prescription for public participation and for consultation with other federal and state agencies (Cortner and Richards, 1980).

Having the advantage of information on resource capabilities, the results of economic analysis, and expressions of public need and other political concerns, professional resource managers must still rely heavily upon their experience and judgment to synthesize these considerations into public land management decisions. I believe that is exactly what Congress intended. The RPA/NFMA planning process represents a net improvement upon the previous system of functional planning. But it should not be confused with a rigid mathematical approach to resource decision making.

Improved Budget Dialogue

A second difference resulting from the planning process is that the RPA Assessment and Program information has enlightened the annual budget dialogue and given it focus. One of the explicit purposes of the RPA was to enhance Congress's role in the overall policy, program, and budgetary process, and to provide it with the facts needed to establish policies and provide consistent program funding. When consistent program funding was available, it was felt, multiple use management would benefit. James Giltmier, for example, has indicated that the principal reason for work on the RPA was to establish a priority work program for the Forest Service so that these priorities would improve the agency's stature at budget time (Giltmier, 1977).

RPA can identify program areas where additional funds are needed. If needed investments are cut back because of national policies, the consequences are identified in terms of lost supply opportunities and increased costs.

The experience of several rounds of budgeting and appropriations confirms that the RPA documents are being used in that way. The RPA Assessment and Program have been used intensively in the Department of Agriculture and the Executive Office of the President, and have provided Congress with much information useful in the appropriate process.

However, those who expected automatic 100 percent funding of the RPA Program adopted by Congress will continue to be disappointed. The Program serves as a guide for budgeting, but plans were never intended to rule the decision; Congress reserves that function for itself. A similar shock awaits those who believe that forest plans will lead to full funding of the work to be done, and that the work will be funded and completed in lockstep with the plan. That simply will not occur.

Congress has come to rely heavily on the RPA documents. The House and Senate committees complained forcefully when the 1980 RPA documents were not available before the budget hearings. In the hearings and markup sessions, committee members typically speak in terms of RPA Program percentages.

The program budgets would have been far less fruitful without the information that RPA has required. Analyzing needs and benefits in nontimber programs and spelling out the range of programs with their costs and benefits in the RPA Program have helped improve the distribution of Forest Service funding over the various program areas.

The value of the Program is that it shows national resource opportunities and displays the associated costs and benefits. However, the needs of the nation and the availability of investment capital will determine the extent to which these opportunities are pursued.

Vehicle Provided for Program and Policy Change
A third major difference that the RPA/NFMA planning process has made is in Forest Service policy and program direction. The process has forced the nation to look systematically at the demands, supplies, costs, and opportunities for its basic forest and range resources. It has prompted an unprecedented degree of serious thought about resource policy issues. This discussion has involved the Forest Service, the Department of Agriculture, presidential policy advisers, and Congress in a very major way.

The process has proved to be a convenient vehicle for periodically reviewing policies and programs to ensure that they are still valid and necessary. That has changed some program emphasis and even prompted some new programs.

The 1975 RPA Assessment, for example, showed the full extent of the untapped potential in Eastern hardwoods, where the growth rate was much greater than the rate of removal. In response the Forest Service initiated an Eastern hardwoods research program that will develop uses for these hardwoods to help offset the demand for softwood timber. One result of this program is press-dry papermaking, developed at the Forest Products

Laboratory in 1980. This process makes it possible to produce strong paper from 100 percent hardwood pulp at reduced cost and with reduced energy inputs. It has many more exciting possibilities.

There is a great deal of flexibility for program and policy change inherent in the RPA/NFMA process. However, expectations that this process would prompt simple but radical change in one or more policies--or that such changes would make vast differences in resource flows--are proving simplistic. Resource policy will continue to evolve incrementally, particularly when one considers the web of resource policies within which we operate and the continuous process of legislative and organizational adaptation to changes in other sectors of society.

Centralized Versus Decentralized Decisions

With regard to a fourth difference since passage of these acts, some have suggested that the most negative impact of the RPA and NFMA on Forest Service management and planning has been in providing centralized direction for an agency that has been traditionally decentralized. There is some fear that this would destroy a strength that has sustained the Forest Service for seventy-five years (Fairfax, 1980).

It is true that the Renewable Resources Planning Act's provision for long-term planning may have centralized Forest Service policy direction. However, the National Forest Management Act has only standardized process and terminology, while leaving decision making in the field. Though the NFMA provides guidelines for forest management, one of the specific considerations in choosing the Humphrey bill over the Randolph bill as a basis for the act was the decision of Congress to maintain the discretion of the professional in the field.

The forest plan and its supporting data bases, analysis, and records provide a foundation for implementing the plan within each ranger district. Now the efforts of Forest Service field managers and their staffs can be concentrated on carrying out the elements of the plan and on monitoring and evaluating the results. This change in focus adds emphasis to carrying out the work in the field. It enables district rangers to update the resource information and to modify the plan to reflect the particular on-site conditions or to address changing needs. Operating within this framework will significantly enhance the effectiveness of managers at the national forest level.

Therefore, there appears to be no significant change in the traditional decentralization of Forest Service operations. In fact, a strengthening of the district ranger's implementation role and the forest supervisor's responsibility for forest-level planning can be expected.

NEEDED CHANGES

The position of the Forest Service in implementing the RPA/NFMA planning process is similar to stepping outside in the morning on the way to work. You cannot completely trust the weatherman's predictions or tell what the entire day will be like from those first few seonds out the door--but you can get a good idea. So you may not set out for work dressed perfectly for the weather ahead, but you can adjust as needed during the day, picking up an umbrella or zipping up your coat. In other words, there is opportunity for change midcourse. The following sections suggest some midcourse changes in the RPA/NFMA planning process.

Streamlining the Process

The first change needed in the RPA/NFMA planning process is to streamline it, making it less complex, less difficult, and less costly to carry out. Its product must be understood and respected by Forest Service managers at all levels, as well as by interested constituents.

The RPA/NFMA planning process was formed in a crucible of conflict. Groups with diverse and opposing views clashed over the many complex issues that arose when Congress considered NFMA. In an effort to resolve conflicting views while maintaining a viable legislative package, Congress created a complex and lengthy planning process. Substance has been submerged in process as a result.

The Forest Service has elected to develop procedures on the run--to produce them while the plans were being developed and while programs were in full operation. This led to forest planning without a complete "end-product" design, particularly for the lead forest plans. Incremental adjustments were made as the planning progressed. With the external critique the draft plans have received, further adjustments have been made in planning procedures.

Some of the best thinking in forestry has gone into these planning procedures in an effort to do the best possible job while satisfying strenuous criticism. Forest Service scientists and outside experts have provided tempting technologies and impressive processes that push Forest Service planners to the outer limits of the applied planning art.

Along the way, a useful planning process has been immersed in layers of legally binding procedure that is complex and difficult to carry out completely. The planning process needs to be streamlined, with compromises made, where necessary, to become more effective in meeting its original purpose and to achieve at least minimum efficiency.

The forest plans must be operational for the forest supervisors and district rangers, as well as being building blocks for national needs. This is a very important need, because failure to satisfy it could diminish the respect that effective plans require and lead ultimately to abandonment of the entire process.

Improved Data Collection and Analysis

The second change that needs to be made is that of taking greater pains in designing, collecting and anlyzing data and in distinguishing the amount of detail appropriate for each planning level. So much detail has gone into some plans that we have approached the limits of management analysis and boggled the minds of reviewers.

Like many large organizations, the Forest Service is grappling to assess and manage the large amount of information that is available to it. Just the right amount of data needs to be collected. In some cases, planning has collected more data than necessary, or even possible, to use.

In addition, more attention needs to be paid to forming the questions today to be answered in the planning process ten years hence. Without asking those questions before designing inventory systems, we cannot expect to collect the data and have the information in shape to address the major frontier issues that planning must address. For example, there is need for better data, better analyzed, on the physical and economic interrelationships among various resource uses. The lack of these kinds of data has hampered analysis as part of RPA planning.

To improve the efficiency of data collection and enhance its accuracy, it is necessary that planning data move along a predictable path. Data collection must be additive so that a data base does not have to be built from scratch every ten years.

There is need for an incremental multiresource inventory system similar to what the Forest Survey has for timber. And there should be universally recognized terms and measurements in this planning system in order to improve the compatibility of data drawn from different sources, and to enhance its utility for planning by other agencies and organizations. The four-agency effort at Fort Collins, Colorado, is addressing these needs.

Equating Resource Trends and Management Programs to Quality of Life
Finally, a third needed change is that of doing a better job of relating resource trends and resource management programs to the larger world. The economic, social, and cultural realities for the nation need to be reflected more clearly in a joining of the present with the future. Resource plans must display opportunities that attract strong public support.

In scheduling investments, resource programs must compete with other claims on the federal budget in a process that is as political as it is analytical. If programs are to merit long-term investments, they must be obviously equated, in the final assessment, to the quality of life for the American people. The nation's resource needs and opportunities make a convincing case for such investments, but their relevance must be shown in the same measure as other claims on the federal budget.

SUMMARY
The comprehensive planning process established by the Renewable Resources Planning Act and the National Forest Management Act has prompted a number of significant changes in Forest Service management and planning. However, since the impacts of these two laws are becoming more evident as time goes on, and since they are interacting with the provisions of other laws, it is still too early to fully assess the extensive effects of this planning process.

My preliminary assessment is that these laws have been a great help in allowing the Forest Service to compete for funds within an austere federal budget. The agency's experience in carrying out this planning process has provided a "shakedown" period, which has revealed several areas where the efficiency and effectiveness of the process can be improved. We remain optimistic that the process is workable.

REFERENCES

Cooper, Arthur E., Robert Wise, and William E. Shands. 1980. "Forest Service Coordination with States and Local Governments," in *A Citizens Guide to the Resources Planning Act and Forest Service Planning* (Washington, D.C., Conservation Foundation).

Cortner, Hanna J., and M. T. Richards. 1980. "Land Management Planning and the Forest Service: Technical Expectations and Political Realization." Draft (Tucson, University of Arizona, and Flagstaff, Northern Arizona University).

Fairfax, Sally K. 1980. "RPA and the Forest Service," in *A Citizens Guide to the Resources Planning Act and Forest Service Planning* (Washington, D.C., Conservation Foundation).

Giltmier, James W. 1977. "Policy Setting Capabilities of the Pluralistic Congress," in Frank J. Convery and Jean E. Davis, eds., *Centers of Influence and U.S. Forest Policy* (Durham, N.C., Duke University).

Hagenstein, Perry R. 1980. "The RPA 1981 Assessment and Program," in *A Citizens Guide to the Resources Planning Act and Forest Service Planning* (Washington, D.C., Conservation Foundation).

Haigh, John A., and John V. Krutilla. 1980. "Clarifying Policy Directives: The Case of National Forest Management," *Policy Analysis* vol. 6, no. 4, pp. 409-439.

U.S. Department of Agriculture, Forest Service. 1980a. *A Recommended Renewable Resources Program*. FS 346 (Washington, D.C.).

———. 1980b. *An Assessment of the Forest and Rangeland Situation in the United States*. FS 345 (Washington, D.C.).

Perry R. Hagenstein

COMMENTS

In chapter 7, Douglas Leisz asks three questions about the impact of the planning requirements of the Forest and Rangeland Renewable Resources Planning Act (RPA) and the National Forest Management Act (NFMA). My comments are directed at his questions, namely:

1. Has the RPA/NFMA planning process brought us closer to national policies for natural resources management?

2. Is the process helping the nation meet its long-range needs for natural resources?

3. Are budgets and major national program decisions markedly better as a result of the planning?

I agree with Leisz's conclusion that it is too early to assess the full impact of RPA. Among other things, we need time to assess how the new national administration is going to respond to national planning for natural resources. But we can reach some tentative conclusions.

National planning has never fared very well in the United States. Some months before his nomination as the 1932 Democratic candidate for the presidency, Franklin Roosevelt said that those unhappy times called for the building of plans. While he was president, Roosevelt created the first, and so far the only, truly multiprogram, nonwartime planning process, the National Resources Board and its various incarnations. It spewed out volumes and volumes of plans, many of which are still classics in their areas. But despite the quality of the plans and the seeming commitment of the president himself to the planning process, they came to naught except as interesting relics of an interesting time. I suppose there are many explanations, but one comes to mind in the context of the present discussion.

The very process of multiresource, multiprogram, multiyear planning runs against the grain of our political way of making program decisions and setting appropriations. Neither the president nor any member of Congress wants to meet his or her constituency on election day with the ability to influence government spending as an elected official seemingly constrained by a bureaucrat's plan.

One must also remember that the other political party has always had little enthusiasm for national planning. Roosevelt's planning was lost in the political battles after World War II. It is my sense that the administration now in office does not believe that the present unhappy times call for the building of plans. Thus, Leisz's cautionary note that the whole story of RPA has not been told is sensitive to the change in political realities that has taken place since the 1980 RPA Assessment and Program were released. Indeed, his suggestions that the whole process be made less complex and less costly is even greater evidence of a sensitivity to political realities.

What sort of standards might be used for judging whether we have moved toward, or away from, national policies for natural resources management as a result of RPA? There is only a very short history on which to base judgments, but three criteria should be considered.

1. The extent of agreement between the president and Congress on the RPA Program;

2. The extent to which the RPA Program "fits" with the other comparable national natural resources policy statements;

3. The extent to which budgets and appropriations for forest and rangeland programs are moving in a coherent fashion with reference to RPA Program goals.

It is my impression with respect to the last of these criteria that Leisz is correct in saying that the RPA Program goals have made some differences in budgets and appropriations. And this, after all, is the clearest expression of national policies for natural resources. The Congress in its appropriations, more so than the president in his budget, seems to be using the Program in its decisions.

My view of the second criterion, however, is quite different from Leisz's view. Not only is there no real "fit" between the RPA and its sister program for soil conservation, but there is no important connection now between the RPA Assessment and Program and the two national programs that are most relevant to it--the nationwide outdoor recreation plan required by the act that created the Bureau of Outdoor Recreation and the nationwide water resources plan required by the Water Resources Planning Act of 1965.

The future of both these plans is now in limbo because of the decision to abolish the Heritage Conservation and Recreation Service (successor to the Bureau of Outdoor Recreation), and because of reported plans to do away with the Water Resources Council. This does not, however, take away from the fact that the 1980 Assessment and Program were not connected to the outdoor recreation and water resources plans then available.

As to the first criterion, it was evident in 1980 that there was little agreement between the president and the Congress on the RPA Program--a point that was not emphasized by Leisz. On this, we can only wait to see what sort of common interests the new administration may find with Congress in the future. I would be greatly surprised, however, if this administration accepts the implicit weights assigned to various resources by the Carter administration.

In sum, there are some signs on which Leisz and I would agree that RPA is leading toward a firmer definition of national policies for natural resources. But I, apparently, am not nearly so sanguine as he says he is about the extent of this movement.

As to the second question--Is the process helping the nation meet its long-range need for natural resources?--the real answer, of course, will be found only at the local level where programs have their effects. Some judgments can be made, however, about the effectiveness of the planning, budgeting, and appropriations process at the national level. Two points are particularly relevant.

One is the extent to which the characterization of national needs for resources has been improved by the Assessment, especially in view of the vast uncertainties about the future. The 1980 Assessment is a substantially improved over its predecessors in the way it describes the interaction of timber demands and supplies. There is some question in my mind, however, as to how much the improved timber analysis is contributing to policy and program formulation. In addition, I wonder if we are well served by a forecast that is conditional on narrow assumptions about future population and economic conditions that are themselves highly uncertain.

In examining future needs, would we be better served with contingency forecasts that predict, in effect, what could happen if there were some major changes in the trend of events? For example, what would be the need for timber if there were a significant disruption in foreign supplies of imported minerals, or if international politics required that the United States greatly increase exports of forest products?

The second point is that more experience with the president's and Congress's response to the RPA Programs is needed before judgment can be made as to whether program levels are responding to needs as defined in the recommended Programs. Actually, the United States has a good record of responding to perceived long-term resource needs. The national forests, the national parks, the Tennessee Valley Authority, the reclamation projects, and more recently the protection of wilderness and wildlife ranges attest to this. But these were responses to broad political concerns rather than to long and detailed analytical reports. If the RPA planning is to assist materially in meeting long-term resource needs, a more effective job will have to be done of translating the national plans into political support for a simple goal or two.

Finally, are budgets and national program decisions markedly better as a result of the planning effort? This question, perhaps more so than the other two discussed above, can only be answered after we have more experience. Leisz notes that the 1980 Program received much attention from the Office of Management and Budget (OMB) with respect to the 1981 and 1982 budget cycles, but this cuts in two directions. For those who want the Program to have an immediate impact on budgets, this may be fine. But the annual budget process has a way of dominating decisions. For those who expected that the program would set standards for judging budgets, as well as appropriations, having the Program ingested into the annual budget cycle is not so fine.

It seems to me that the notion that the five-year Program would contain purely professional analyses of Forest Service program needs, against which both budgets and appropriations coud be judged, was never very plausible. Government just does not work that way. But to go on from here and accept the Program as just another piece of information for those who make budgets and appropriations will ultimately lead to the Program's being an exercise of little importance. As with the 1980 Program, the particular budget concerns at the time decisions on the Program are made will dominate the recommended Program. The effectiveness of the Program in improving budgets and appropriations will depend in large part on the ability of the Forest Service, and the willingness of OMB and the Congress, to keep the Program independent of the budget process, but relevant to budget decisions.

Chapter 8

NATIONAL FOREST PLANNING: AN ECONOMIC CRITIQUE

John L. Walker

The purpose of this chapter is to provide a critique of the economics involved in National Forest Management Act (NFMA) planning. This is a very difficult assignment because, while formal direction for NFMA planning is gradually emerging in the form of interim directives, agency memorandums, and draft reports, the whole process is still in a very active state of development.

Ever since the Monongahela decision (<u>West Virginia Division of the Izaak Walton League</u> v. <u>Butz</u>, 1975) which triggered the legislative process that produced the National Forest Management Act of 1976, I have followed the developments in Forest Service planning as closely as possible in order to understand and influence the implementation of what I regard as the most important act affecting national forest management. My interest is in seeing as much sound economics as possible incorporated in Forest Service planning.

The Forest Service's historic objective of maximum sustained yield of timber with some multiple uses added in appears to have shifted to maximizing some very poorly defined measure of environmental quality, including the provision of more dispersed recreation opportunities. The historic preoccupation with sustained yield and several related concepts--such as regulation of forests to obtain an even distribution of age classes, rotation ages that maximize mean annual increment, and evenflow, which now has a more rigorous definition and the new label of nondeclining yield--continue to dominate the planning process to the detriment of both sound timber management and environmental quality. The NFMA planning process I see emerging slowly, painfully, and expensively is badly flawed by the imposition of management constraints on forest land use programs with little or no concern about the resultant benefits or costs to society. The basic cause of this phenomenon is not environmentalism, which has simply become the current scapegoat. It is the lack of rational economic objectives. Without a sound economic framework for land management planning, society's resource output opportunities are not made visible, resource tradeoffs are not accurately measured, and management decisions are not made and justified in a rational manner.

The forestry profession in the United States and elsewhere in the world was founded on a rejection of markets to allocate resources and on the belief that free enterprise would not perpetuate timber supplies to meet future demands for timber. Consequently, there is a long tradition in forestry, particularly in public forestry, of plans that use a materials balance approach. Planning has been done without the use or understanding of the key role interest rates and prices play in making market economics operate efficiently. This historic perspective should be kept in mind throughout the following discussion of the planning process emerging today for implementation of NFMA.

SIGNIFICANT LEGISLATION

The National Forest Management Act of 1976 is very long and complex. Most of it serves to enlarge the Forest and Rangeland Renewable Resources Planning Act of 1974, commonly referred to as the Resources Planning Act, or simply RPA, to almost three times its original length. The basic thrust of NFMA and the amended RPA is more planning. RPA planning will continue at the national level with periodic assessments of the overall forest resource supply and demand situation and periodic reviews and updating of the program for the Forest Service. RPA Assessments were completed in 1975 and 1980 and subsequently are to be updated every ten years. RPA Programs were also completed in 1975 and 1980, with updates to continue on a five-year cycle.

The NFMA directed the Secretary of Agriculture to promulgate final regulations within two years of the enactment of NFMA, or by October 22, 1978. The process of developing these regulations was so complex and controversial that they were not finalized until September 17, 1979, almost one year late. Even then resolution was deferred on some of the more controversial issues that the act specifically required the regulations to address. The regulations accomplished this deferral by directing that the chief of the Forest Service establish economic analysis guidelines to become effective within one year from the date the regulations were issued. Complete, comprehensive guidelines are still unavailable, although the Forest Service published Interim Directive No. 7, <u>Economic and Social Analysis</u> (USDA, Forest Service, 1980f), on September 2, 1980. It might be argued by the Forest Service that this directive now meets the legal requirement of the NFMA regulations. Interim Directive No. 7 provides some general policies and principles, but more specific direction is needed to clarify how those policies and principles are to be applied. The real economic guidelines are still in draft form, but it would be improper for me to base my critique on unofficial material that is still subject to change. Instead, my critique is based on other documents that have emerged from various sources within the Forest Service, and thereby perhaps influence what eventually emerges in the final official Economic Handbook and, more importantly, in the final forest plans, as will be discussed later.

It should be emphasized that the NFMA is not a prescriptive act. It gives the Secretary of Agriculture and the Forest Service a great deal of flexibility. There are some troublesome provisions, such as the requirements that "prior to harvest, stands of trees throughout the National Forest System shall generally have reached the culmination of mean annual increment of growth" (16 U.S.C.A. 1604 (M)(1)), and that "the Secretary of Agriculture shall limit the sale of timber from each national forest to a quantity equal to or less than a quantity which can be removed from such forest annually in perpetuity on a sustained yield basis" (16 U.S.C.A. 1604 (e)(1)). These raise the zero-interest approach to timber harvest age calculations and evenflow nondeclining yield issues that are irrational from an economic perspective. The allowable cut effect also comes up once or twice in the act. The more controversial sections were written in sufficiently ambiguous language, however, to achieve successful compromises among sharply differing interest groups: at the time the act was passed, both the Sierra Club and the National Forest Products Association claimed a victory. But taken as a whole, the act provides numerous requirements for economic analyses, as has been pointed out by Krutilla and Haigh (1978) and is emphasized again by Krutilla, Bowes, and Wilman in chapter 6 of this volume. The ambiguity in the act simply shifted the conflict first to the regulation-writing process and next to the actual planning process being undertaken at the regional level and individual forest level. The act requires each national forest to complete a comprehensive land and resource management plan by September 30, 1985, and then to update it at least every ten years. The NFMA regulations added a requirement for regional plans.

In spite of many ambiguities and troublesome inconsistencies, the NFMA will permit the Forest Service to take almost any action it can reasonably justify. Sound economic analyses must play a central role in the justification process. In this area the act is not ambiguous. As I wrote earlier (Walker, 1977), the administrative discretion permitted by the act allows most provisions to be interpreted in a manner much more consistent with economic efficiency than is the case with current Forest Service policies and procedures. The regulations promulgated in September 1979 actually strengthen the requirements for economic efficiency in national forest management. Even the controversial nondeclining evenflow provision is not necessarily inconsistent with economic efficiency. There is a specific departure provision so that "in order to meet overall multiple-use objectives, the Secretary may establish an allowable sale quantity for any decade which departs from the projected long-term average sale quantity that would otherwise be established." The only requirement for such a departure is that it be planned to "be consistent with the multiple-use management objectives of the land management plan" and that "plans for variations in the allowable sale quantity must be made with public participation as required by section 6(d) of this act" (16 U.S.C.A. 1604(e)(1)).

FOREST SERVICE AUTONOMY

Most of this chapter will discuss the economic implications of NFMA implementation, but first some comments about Forest Service autonomy. As mentioned above, all that the act and subsequent regulations really call for is that a planning process that utilizes public participation be developed and followed, and that economic and environmental criteria be considered in order to develop balanced programs. The Forest Service apparently views this broad discretion as a justification for business as usual, and has interpreted the act so as to reinforce current policies and procedures. In fact, I believe one of the reasons for the fact that much of the required NFMA direction, such as the economic analysis guidelines, is still unavailable is so that the actual planning based on informal direction can proceed beyond the point where significant changes can be made. Critics will then be told that their ideas will have to wait for the next round of planning before they can even be considered for adoption. As has happened in the past, it will be pointed out that all the forests must have their new plans completed by a certain deadline--this time it will be September 30, 1985--and there simply will not be time to go back and start over again and still meet this schedule. Tight schedules and the lack of time for substantial revision become the ultimate defense for business as usual.

I realize that changing the direction of a long-established federal bureaucracy is somewhat like steering a supertanker. You can't just give the rudder a sharp turn. The situation has become so serious, however, that we cannot wait until 1990, or even 1985. Somehow, we have to pick up the ship and put it back down heading in almost the opposite direction.

President Carter's Departure Directive

In April 1978 when President Carter was casting about for ways to stop inflation, he instructed the departments of Agriculture and Interior, the Council on Environmental Quality (CEQ), and his economic advisers to report to him within thirty days on how to expand timber harvests. The study team reached a stalemate over what to recommend, and no report was ever produced, although some good advice did get through to the president from his economic advisers. On July 19, 1979, over strong objections from the Forest Service and CEQ, President Carter formally directed the secretaries of Agriculture and Interior to "use maximum speed in updating land management plans with the objective of increasing the harvest of mature timber through departures from the current non-declining evenflow policy"

(Federal Register Office, 1979, p. 1294). This direction was reiterated in the president's 1980 RPA statement of policy that accompanied the 1980 RPA assessment and program submitted to Congress. The statement of policy emphasized that

> this nation's housing requirements during the next five years [not by 1985 or 1990] are expected to place major demands on the forest products industry to increase production of lumber and wood products. In the long run, private forestlands are expected to become a chief source of increased timber supply, but <u>during this decade</u> [emphasis added], careful consideration must be given to increased supplies from federally-owned lands, particularly the National Forest System. The recommended program for the National Forests provides for timber harvest goals ranging between 11.0 and 12.5 billion board feet by 1985, based on traditional planning guidelines. However, at my direction, the Secretary of Agriculture is accelerating national forest land management plans with the objective of increasing the harvest of mature timber through departures from the current non-declining evenflow policy (USDA, Forest Service, 1980a).

The traditional planning guidelines referred to by the president include nondeclining evenflow.

At the Washington office level, lip service was paid to the president's directive. Forests were designated for accelerated planning with completion dates moved up to 1981 and 1982. At the regional and national forest levels, however, the directive is being ignored, and not without the knowledge and active consent of the Washington office. It has been suggested to me by John McGuire, retired chief of the Forest Service, that the president did not really understand what he asked for and would not have wanted it if he did, and that the directive simply provided something that looked like it would help fight inflation. I must admit that the directive was certainly out of character with most of the related actions of Carter's administration and that the president did not follow through and demand an appropriate response to his directive from either the Forest Service or the Bureau of Land Management.

NFMA IMPLEMENTATION

This section will outline what seems to be happening with NFMA planning at the individual forest and regional levels. As already noted, this planning is proceeding without the benefit of all the formal direction required by NFMA and subsequent regulations. The economic analysis guidelines and other important Forest Service manual changes have been unavailable or are incomplete. Instead of the normal rule-making process, there have been directives in the form of in-service drafts, interim directives, and other forms of in-service communications that are not subject to legally required public review and comment. From time to time, when various critics in the industry have been told about what is going on and even given copies of some draft material, they have offered informal comments and pointed out needed changes or inconsistencies among regions and forests. The response has been comments to the effect that "It's still unofficial," or "It's been changed," and so we are left to put the pieces of the puzzle together as best we can.

Forest and Regional Plans

In the meantime, every national forest and region now has a planning team in place. All have scheduled completion dates for their plans. They are proceeding by making many important decisions as necessary on an ad hoc basis. In all fairness to the people on these planning teams, most are highly dedicated and conscientious individuals. They are trying to do the best job they can, given the time available, lack of specific direction, and inherent policy and planning biases. Many forests and some regions have circulated documents on so-called decision criteria, planning goals, planning alternatives, analyses of the management situation, and so forth. These are all based on planning steps required by the regulations.

One forest, the Lolo National Forest in western Montana, has actually published its draft environmental impact statement (EIS) and proposed forest plan. This was in April 1980, with a period for formal public review and comment that was extended to September 1980. In November 1980, Region One released its draft EIS and proposed region plan. I participated on industry review teams that analyzed the Lolo and Region One plans. Although my company has no operations in Montana, we believe that getting these precedent-setting plans off on the right track is the only way we can effectively influence the planning that will directly affect us. These draft plans are the only hard evidence we have on the kind of product the Forest Service may deliver. We cannot wait a year or two when plans will be popping out all over the country. It should be noted that, when added together, all these new plans will make up the 1985 RPA Program. If they are not done right, we will be caught in a vicious circle of plans determining the next RPA Program and the RPA Program driving the next set of plans. Before this golden age of planning arrives, something must be done.

Preliminary material on decision criteria, goals, objectives, and alternatives is so steeped in ambiguous "planese" that it is meaningless. The draft Lolo EIS and forest plan are totally unacceptable. Many fundamental changes need to be made. I was very gratified to learn that the Forest Service has decided to redraft the Lolo plan, although it is not clear yet what the agency plans to do with the draft Region One plan. Finalized plans in substantially the same form as either the draft Lolo or draft Region One plan would unquestionably lead to appeals and litigation. Substantially changed final plans without redrafted plans and EIS would also be very vulnerable to successful legal challenges.

The Lolo is one of the forests selected for accelerated planning as a result of President Carter's departure directive. Departures were considered in a very short, two-sentence paragraph buried deep in the draft EIS. This paragraph simply said that a departure alternative had been considered and rejected as unacceptable. The departure alternative was not even described or included as a formal planning alternative. Further investigation by the industry review team revealed that one computer run called the departure alternative had been made using assumptions of such nature that the results were unacceptable.

The interdisciplinary teams assembled to prepare NFMA plans consist of numerous resource specialists. Rather than providing information on their particular resource and on its interactions with other resources so that the information can be used in a rational decision-making process, these specialists are apparently negotiating and specifying numerous management constraints at the staff level. The responsible official for each plan can only be vaguely aware of the impact of the numerous decisions that shaped the EIS and proposed plan before it reached that official's desk for approval or disapproval. Most important, multiple use and land allocation decisions are actually being made by planning teams before they even formulate plan alternatives. Consequently, the range of alternatives is unduly constrained, and resource tradeoffs and opportunity costs are impossible to measure properly.

The planners are all too often simply playing word games in "planese" when they define decision criteria, goals, and plan alternatives. The draft Lolo EIS and plan are a good example: one criterion is that a plan be responsive to the needs of local people; thus a plan alternative was developed that supposedly best meets this criterion. Another criterion is that of maintaining "a diverse mosaic of vegetational development well distributed across the forest" (USDA, Forest Service, 1980e). The comparable plan alternative is specified so that "a variety of plant and animal communities in balance with their physical environment comprise the forest" and "the principal goal of management is both in the way the forest looks and in the way it functions" (USDA, Forest Service, 1980d). The recommended plan that somehow emerges from all this is simply asserted to be the best for meeting all the conflicting criteria and goals. Thus the planning process is not being used to develop meaningful alternatives that can be measured against objective criteria to arrive at sound decisions. All the Lolo plan alternatives that supposedly respond to such divergent criteria as achieving economic efficiency, maintaining natural environments, or emphasizing service outputs are so constrained that they represent only minor variations of the same basic plan, which the planning team in its collective wisdom offered up as the preferred alternative.

Another major difficulty with NFMA planning at the forest level is the computer program that is supposed to make it work. An extremely and unnecessarily complex model called FORPLAN is being used by the forest planning teams while it is still being developed and debugged. Numerous errors have already been found in the model. It simply will not be possible to test and document properly the many features of this model before the first cycle of planning is supposed to be complete.

The regional NFMA plans apparently will provide regionalized silvicultural guidelines and other direction to individual national forests. They are intended to provide a link between the national-level RPA Program and individual forest programs. Regional plans are not called for in the NFMA, but are called for by implementing regulations. There is very little substance in the Region One draft EIS (USDA, Forest Service, 1980g) and proposed plan (USDA, Forest Service, 1980h). The required guidelines for silvicultural systems, size of clear-cuts, and so forth, seem to avoid controversial standards. The approach was to list all kinds of very general alternatives and to leave to the individual forests the final decisions that would be based on criteria that best meet their local resource management objectives. In general, the draft EIS and proposed plan contain generalizations about issues and concerns, highly arbitrary allocations of the region's 1980 RPA goals for various resource outputs to different alternatives, and highly exaggerated estimates of the money and personnel required for each alternative. These regional plan documents contain none of the economic analyses required by the National Environmental Policy Act (NEPA), NFMA, and the regulations.

1980 RPA PROGRAM

The 1980 RPA documents, particularly the Program (USDA, Forest Service, 1980c), are also unacceptable. The Program output goals are very inconsistent with the findings of the Assessment (USDA, Forest Service, 1980a). The 1980s, for example, are projected in the Assessment to have an unprecedented increase in the demand for timber, caused largely by the postwar baby boom population moving into its home-buying years. The only timber supplies available to meet this unprecedented demand are acknowledged to be on the national forests, yet the Program establishes timber goals that would result in lower national forest timber production. The only way to achieve higher timber output would be through departure alternatives developed at the individual forest level and approved by the chief of the Forest Service.

The 1980 RPA output goals deserve some discussion. The RPA documents were delivered to Congress several months late because of strong differences between Forest Service and Office of Management and Budget (OMB) officials. To avoid having the unfinalized documents subpoenaed for hearings, the two agencies patched together a compromise Program that could be subject to further resolution under the RPA statement of policy and annual budget requests.

Regarding the dispute between the two agencies, evidently OMB does not believe that the Forest Service provides adequate analyses for any of its programs. With respect to timber sales, OMB was not willing to accept the proposed Forest Service goals as worthy of the name. OMB felt there were too many important policy decisions undergoing review that needed to be resolved. These included the RARE II (Roadless Area Review) impacts on the land base available for timber production, the nondeclining evenflow policy, and the multitude of environmental and other multiple use constraints that impact timber management.

For example, the Forest Service proposed a timber goal of 12.5 billion board feet (bd-ft) for 1985, together with substantial funding and personnel increases needed to achieve it under their current policies. OMB wanted to provide a range in which they estimated the outcome of the policy decisions would fall. The low OMB goal for 1985 was 11.0 billion bd-ft, which assumed no increase in funding and personnel. The high OMB timber goal was 17.0 billion bd-ft, which assumed some departures from nondeclining evenflow. In the scramble to get a compromise program together, the high OMB goal was eliminated in favor of the departure language in the president's statement of policy. The high RPA goal became the Forest Service's original 12.5 billion bd-ft proposal.

Unfortunately, many in Congress have put far too much emphasis on the range of output goals in the RPA. They wanted the Forest Service's best professional judgment of what it thought its program should be. One of the last acts of the 96th Congress was to revise the RPA statement of policy to accept generally the so-called high goals in the Program, with the qualification that, with respect to timber, range, and even watershed, the high RPA goals are not enough. The new statement of policy emphasizes that all forest and rangelands should be managed "to maximize their net social and economic contributions" (Congressional Record, 1980). Not enough attention was given by Congress, however, to the flawed premises that underlie the entire RPA Program. Somehow the reference to departures was also dropped.

FOREST SERVICE BUDGETS

I believe that OMB was acting responsibly to help bring necessary reform to Forest Service programs. The shortsighted accountant mentality attributed to OMB is based on a poor understanding of what that agency wanted to do. I think the professional staff at OMB includes a handful of people in Washington, D.C., who really understand the fundamental reform needed in federal timber management. OMB Director David Stockman may someday have a double-bitted axe to hang on his wall. One blade should be labeled "timber," and the other, "budgets." There is ample opportunity to cut more of both.

OMB has a very pivotal role in reviewing Forest Service budgets, which, for more than two decades, have been very controversial. The Forest Service consistently wants bigger budgets. OMB generally reduces the amounts requested by the Forest Service for the budget the president submits to Congress. Congress usually appropriates more funds for the Forest Service than those requested by the president. The forest products industry and other user groups lobby intensely for bigger appropriations.

With respect to timber management, the Forest Service tells industry that if timber sales are to be increased, it needs more money and personnel, not just for timber sale preparation, but for reforestation, timber stand improvement, and the like. In fact, the Forest Service has frequently argued that to achieve balanced programs under RPA it also needs more funding for recreation, wildlife, and so forth, before it should cut more timber. The Forest Service is holding its overmature old-growth timber hostage to increased appropriations. Under the nondeclining evenflow policy, the allowable cut effect from growing young stands of trees faster is the only way to cut more old-growth timber. OMB does not accept this, rightfully so, and wants to force a reconsideration of the evenflow policy rather than just accepting the policy and providing more money.

The national forest system currently costs U.S. taxpayers about $1 billion per year. Receipts for FY 1980 were a little over $1 billion while the authorized budget was $2.2 billion. The budget does not include any interest charges on the tremendous capital asset that the national forests represent. If this asset were in private ownership, it would be returning $5 billion to $6 billion per year to its owners rather than requiring a billion-dollar subsidy. This comparison should illustrate the magnitude of the need for change. I am very hopeful that the new Congress and administration will provide some badly needed leadership.

One final topic related to Forest Service budgets needs mention before discussion of specific NFMA planning issues that need to be resolved. The FY 1981 appropriations bill (HR 7724) had a House-approved timber sale level of 12.2 billion bd-ft. The Senate appropriations committee cut this amount to 11.9 billion bd-ft. The committee reduced the timber sale level because it was concerned that 500 million bd-ft sold in FY 1979 and an anticipated 1 billion bd-ft sold in FY 1980 are going uncut. The committee said this implies that timber is being held speculatively. It was apparently unaware that there is normally from two to two and half years of uncut timber sale volume under contract. This is 20 billion to 25 billion bd-ft, not the paltry 1.5 billion bd-ft referred to in the Senate appropriations committee's report.

In my opinion, much of the concern about speculative bidding is misplaced. Several Forest Service proposals have surfaced recently to deal with this "problem." The Forest Service's thinking on how to respond to the perceived speculative bidding problem is just the reverse of what it should be. They think in terms of further regulating and restricting markets. I strongly believe the speculative bidding "problem" would disappear if the Forest Service would help create an open market for timber by increasing the volume under contract to coincide more with the market cycle and optimal inventory levels for large, modern, processing facilities. A 4- to 5-year supply under contract would be much better than the current 2.0- to 2.5-year supply. If timber sale contracts were easily transferable, speculators would become a stabilizing force in the market. In well-organized competitive markets, speculators become arbitragers who buy and sell to profit from price differences, and thereby tend to equalize prices. If mill owners had the option of purchasing existing contracts rather than depending primarily on creating new contracts by bidding on Forest Service sales, they would have a far more stable and predictable timber supply. Speculators would help ensure that a ready market for harvestable timber existed and would bear the costs of holding timber under contract, provided there was a sufficient total volume under contract to make the system work.

Virtually all the Forest Service ideas seem designed to regulate timber cutting further in hopes of achieving results that are closer to the idealized evenflow model. The forest industry, as well as local, regional, and national economies, would be far better served by a system that let free market forces price and allocate public timber.

NFMA PLANNING ISSUES

For any process as complex as NFMA planning, numerous issues could be discussed. The following will focus on just a few issues that I believe have highly significant economic implications and are directly related to what I believe are the necessary steps for developing economically optimal plans. The use of "economically optimal" here does not mean just maximizing financial returns from timber sales, but includes allocating all the forest resources over time in the manner that maximizes their net contribution to social welfare. Planning methodology that uses correct economic criteria, together with a management system that is flexible enough continually to subject all the planning assumptions to reality checks and make the necessary adjustments, can result in both economically optimal plans as well as performance. Without such methodology and a flexible management system, all that can ever be expected from NFMA plans are nonsensical analyses or undocumented assertions that, in the judgment of the planners, the recommended plans somehow achieve optimal results.

Properly documented plans will require a methodology that makes visible all the key planning assumptions and provides an economically rational measure of all the costs and benefits associated with each significant decision embodied in each plan. The way to start is with an analytical framework designed to maximize the net present value of all future costs and benefits. The interest rate used and the measurement of the costs and benefits require careful analyses. Both monetary and nonmonetary costs and benefits need explicit consideration. In these areas I find the emerging NFMA process to be very deficient.

The main deficiency is caused by the resistance of Forest Service planners to using economics. This resistance is reinforced by the various types of direction they are receiving from the regional and Washington offices. Rather than starting with a zero-based planning process and an economically rational methodology to justify each element of what emerges as the recommended plan, it appears that most planning teams are starting the process with many preconceived management constraints and are unwilling to consider the costs and benefits of these constraints. The major constraint is that of nondeclining yield. Other constraints include the determination of the land base available for various resource uses, rotation ages for timber, ending inventory, and a whole host of management prescriptions. The draft Lolo EIS and plan that have been withdrawn are, it is hoped, the worst case examples. The various alternatives displayed in these drafts were so constrained that widely different philosophies supposedly represented by different alternatives were in fact minor variations of the same alternative.

Demand Analyses

The approach that I have proposed to Forest Service officials on several levels is to start with a sound analysis of the markets for Forest Service stumpage. This includes determining appropriate market areas, the current and prospective demand for timber within each market area, the supply of non-Forest Service timber, and then translating these into demand curves for Forest Service timber. Such an approach, at least in the West where national forests frequently contain 50 percent to almost 100 percent of the standing timber inventories within a market area, will unquestionably lead to the conclusion that the quantity of timber sold by the Forest Service during any given time period, and how it is sold, does have a marked effect on the price of stumpage and delivered logs. Even when the mills that manufacture lumber and plywood from this timber sell into competitive national and international markets where their output cannot affect prices, the relatively fixed mill capacity in a timbershed during any short time period, such as a year, gives the Forest Service monopolistic power over the mill owners. In economic texts this kind of market situation is called an oligopoly. Some oligopolies are characterized by one giant firm and a number of smaller ones. The Forest Service, as the dominant owner of timber, frequently fits the textbook example of the giant firm. The theory for price and output decisions in such markets is well developed and is usually based on the assumption that the dominant firm will set the price. The smaller firms then take this price as given and produce that output at which their marginal costs equal the price set by the dominant firm. The Forest Service reverses this process by setting its output through an allowable cut calculation and then tries to ignore the price effects of its action.

In the 1980 draft of the forthcoming economics handbook, the Forest Service acknowledges it can be argued that

> in many cases, the Forest Service does comprise a significant portion of the market for many of its forest outputs. An assumption of perfectly elastic demand applying to the forest would, in such cases, appear unjustified as changes in the level of forest output would be expected to influence market prices (USDA, Forest Service, 1980b).

The draft handbook goes on, however, to assert that

> assembling information that captures the elasticity of a demand schedule at the forest level would be a heroic effort at best and frustrating in any event. In addition, explicit recognition of market dominance and inelastic demand raises the real question of pricing strategy--an approach that is rejected out-of-hand since the Forest Service does not operate as a profit-maximizing producer (USDA, Forest Service, 1980b).

The draft handbook concludes that

> a simple procedure that is recommended in the absence of more definitive information is to assume that the demand for forest products at the estimated benefit level is perfectly elastic over an established quantity interval. . . . The use of a cut-off point on an elastic demand curve is to implicitly recognize the existence of price elasticity without requiring extensive mathematical rigour for determining it. . . . Estimating the interval over which the unit benefit applies is the real crux of this procedure. The objective, in most applications is to estimate the quantity in excess of which output can be considered to have no value through use or consumption. . . . Demand information must be projected over the planning period. Using this simple procedure, an entire demand schedule would not have to be projected, but rather only the extreme point of the quantity interval and the applicable benefit value (USDA, Forest Service, 1980b).

I have omitted the references to procedures on how to estimate the so-called unit benefits and quantity intervals, since they only obfuscate the intent of the handbook. The recommended simple procedure is simply a sham. Unless the NFMA planning process can be headed down a much different track, all economic analyses associated with it will be no more than pretense. But if economics is not used to make any decisions, perhaps pretense is all that is needed.

The fact that the FORPLAN computer program is far from being debugged presents a very real problem. The option to use downward sloping demand curves is one feature of FORPLAN that does not work even when demand curves are available. Some of the test runs that Forest Service planners have made at industry's request indicate that FORPLAN may in fact be constraint-dependent. When the program is used with so many constraints on the solution that it is in effect a simulation rather than an optimization model, it can produce reasonable-looking results. All attempts to obtain unconstrained solutions have been very unsuccessful. The FORPLAN program is so complex, however, that the Fort Collins computer center becomes swamped when several forests are attempting to use the program. It has become necessary, I understand, to limit severely the number of forests that can use the program at the same time and also to limit the number of runs each of these forests can attempt to make each day. Adding the complication of downward sloping demand curves greatly increases the computer time needed for each run, since FORPLAN uses a very cumbersome, computationally inefficient approach. A stepwise approximation technique is used with the downward sloping demand curves to obtain linear programming solutions for what is really a nonlinear problem. A February 3, 1981, memorandum from Norman E. Gould, Director of Timber Management at the Washington office, to all regional foresters provides a very practical solution to this problem. Gould said that "at this time, we feel that there is insufficient information to build downward sloping demand curves for individual forests. We recommend that you not attempt to build and use them for this reason." Gould referred to a research project on this subject at the Pacific Northwest Station in Portland. Based on what the researchers say, it is quite clear that their research schedule will provide nothing for the forest planners that need help now.

This research has already produced some tentative regional demand curves for timber that are linked to the 1980 RPA Assessment. A draft paper by Haynes, Connaughton, and Adams presents some coefficients for downward sloping demand curves for timber, but assumes that whatever prices prevail on a regional level must be taken as given by individual national forests. In a draft on Region 2 timber demand analysis, Benninghoff and coauthors also assert that "in most cases the demand curve for stumpage from an individual Forest can be assumed to be horizontal. That is to say, harvest levels on a Forest are not likely to vary enough to influence stumpage price" (Benninghoff and coauthors, 1980). The inventory available on many individual forests, however, is more than sufficient to change regional prices, and it would seem logical that prices in the vicinity of an individual forest would be changed more than regional prices if that forest changed the amount of timber it sold sufficiently to change regional prices. The key assumption, apparently, is that individual forest harvests will not be allowed to vary enough to affect prices, and therefore the demand curves can be treated as if they were horizontal.

One of my contributions to industry's response to the draft Lolo plan and EIS was to develop a demand curve for Lolo National Forest timber. I used published information on log flows to define the Lolo National Forest market area. I then used mill capacity, logging cost, and log price data to demonstrate how to develop demand curves that will estimate the impact Forest Service timber output will have on prices. Others in the industry as well as myself have urged the Forest Service to develop such curves for all national forests and to use them in the FORPLAN model to develop optimal plans. Only a handful of people in the Forest Service are even mildly responsive to this approach. It appears that the Sierra and Six Rivers national forests in California may try to use this approach, but most Forest Service planners are strongly opposed to it and view it as an expensive, irrelevant task that will make their work unnecessarily complex.

The favored approach within the Forest Service is to pick a set of prices over time relating each forest's historic average prices to projected regional price trends that are based on the 1980 RPA Assessment. Each forest would then assume these prices would prevail over time regardless of their level of output. They defend their position by arguing that over the limited range of timber outputs that they expect to analyze, it would be impossible to make meaningful distinctions in price. With my Lolo demand curve, for example, they point out that prices vary from only $73.10 per thousand bd-ft to $68.65 per thousand bd-ft as the output of their alternative varies from a low of 95 million bd-ft to a high of 135 million bd-ft per year. This narrow range of output is due to nondeclining yield harvest flow constraints. If the Lolo planners really believed their $73 per thousand bd-ft would prevail regardless of the level of output and removed all constraints from the FORPLAN model, the solution would be to cut the entire 10-billion bd-ft inventory on the Lolo National Forest in the next decade, since decades are the smallest discrete time periods used for planning. If years were used instead of decades, the unconstrained solution would be to cut the entire inventory in one year! This assumes that the rate of discount being used is higher than the projected rate of price appreciation. If the projected rate of price appreciation were higher than the discount rate, the unconstrained solution would be to cut no timber that time period.

This may sound like nonsense, and it may well be, but not because of the lack of harvest flow constraints. If cutting all or none of the inventory on one national forest in a given year or decade will not affect prices, then that national forest is not more important to the producers and other suppliers within its market area than is Farmer Joe's woodlot. Mills would continue to buy their raw material, and other stumpage producers would continue to sell their output at the unaffected price level regardless of the actions of the Forest Service. If this were really the case, I submit that there would be so little interest in national forest management that volumes like this would not need to be published. All the national forests in the West with significant inventories or commercial timberlands can and do affect the prices of stumpage and delivered logs.

Cost-Benefit Analyses
If properly done, NFMA planning should be nothing more than a series of cost-benefit analyses that lead to economically optimal forest plans. There is a large literature on cost-benefit analyses, but it has largely been ignored by Forest Service planners. This is unfortunate, since there is no need to reinvent many of the basic concepts and procedures that should be used. Much detailed methodology needs to be developed and standardized for NFMA planning, but it should all be based on accepted economic theory for cost-benefit analyses. NFMA planning, based on sound economic theory, can and should accommodate resource uses that have both monetary and nonmonetary values.

The approach industry has suggested to the Forest Service is to start the alternative formulation process by making an unconstrained benchmark run for each forest. This unconstrained run would include all the resource outputs for which monetary values can be properly handled by the FORPLAN program. At a minimum this must include timber, with the capability to handle downward sloping demand curves functioning correctly. Other resource outputs that could logically be included in unconstrained benchmark runs are recreation, grazing, and water. This would require the use of downward sloping demand curves when appropriate for these resources. FORPLAN, however, is not designed with this capability for resources other than timber. Unconstrained benchmark runs would then be used to determine all the resource outputs over time that are associated with the maximum net present value for those resources whose monetary values were explicitly and correctly entered into the FORPLAN objective function.

After inspecting an unconstrained benchmark run, Forest Service planners can then prescribe a set of constraints that might logically be imposed on the unconstrained solution to account for various nonmonetary values. Since there undoubtedly will be a lot of interaction among these constraints, it will be necessary to develop a rational methodology for sequencing their imposition on the unconstrained solution in order to obtain valid measures of their opportunity costs. It will be impossible to make separate computer runs for every possible combination of a set of constraints. This would require 2^n runs, where n is the number of constraints. Five constraints would require 32 runs. Ten constraints would require 1,024 runs. Twenty constraints would require 1,048,576 runs for an exhaustive enumeration of all possible combinations. Exhaustive enumeration is simply impossible and, I think, unnecessary for NFMA planning, even though a large number of constraints will end up in every recommended plan. It should be possible to impose each significant constraint singly on the unconstrained solution and to obtain a measure of its opportunity cost by observing the reduction in net present value. Planners could then be expected to provide well-reasoned arguments as to what nonmonetary values can be obtained by the imposition of each constraint, and why in their judgment some of these nonmonetary values may exceed the monetary opportunity costs. When nonmonetary opportunity costs can be identified, they should also be explicitly considered. The public would then have the opportunity to agree or disagree with these arguments as part of its participation in the planning process. The planners' arguments and the public response would then be available to the Forest Service official who must make the final decision.

After the individual opportunity costs of various constraints have been measured and have identified benefits that exceed these opportunity costs, the planners can use their judgment to decide which constraints do not interact. The constraints that do not interact can then be grouped all together for another run, which will become a new benchmark. Interactive constraints can then be imposed on this new benchmark run with the planners developing explicit arguments about incremental nonmonetary benefits associated with incremental opportunity costs. This process should continue until a limited number of alternatives would emerge for inclusion in a draft EIS. A draft forest plan would include the alternative that maximizes the net social benefit from all resource uses as justified by the planners and agreed to by the responsible Forest Service official. The other alternatives in the EIS would include different sets of constraints to reflect different judgments about whether or not various nonmonetary benefits exceed the opportunity costs.

It would be possible and useful to take the set of constraints associated with the recommended plan and perform a sensitivity analysis. This could be done by lifting each constraint one at a time to measure its opportunity cost when interacting with all the other recommended constraints. This process would enable the planners to modify some constraints in order to "fine tune" the recommended plan. It has been suggested in some of the draft manual and handbook material that this sensitivity analysis based on a recommended plan is all that is necessary. I strongly disagree, since the planners, the public, and the decision makers would not have any rational process for constructing, testing, and justifying each assumption and constraint that was in such a plan.

The interaction of all the constraints in a recommended plan can result in the apparent opportunity costs being grossly in error. The use of "shadow prices" associated with standard linear programming applications and knowledge of how much a particular binding constraint can be relaxed before another constraint becomes binding is simply not an appropriate way to attempt to measure opportunity costs for NFMA planning. There will be far too much interaction among the constraints in recommended plans, given the current state of the art. The use of FORPLAN will depend heavily on constraint specification rather than on the specification of production functions and marginal costs and benefits in a way that would permit the objective function of FORPLAN to produce optimized tradeoffs as part of the solution to a single computer run.

Significant Constraints

There are almost unlimited numbers of management constraints that can be included in a final forest plan. Some of these would be very explicit. Others would be buried in the many assumptions used to develop yield tables and management prescriptions that FORPLAN selects from when it "optimizes," subject to the explicit constraints. All of these constraints can be important. If the interdisciplinary teams responsible for formulating management prescriptions and constraints were guided by economic criteria, it would be far more likely that the prescriptions and constraints would be consistent with the objective of maximizing the present value of net social benefits. This is another facet of the problem caused by the lack of sound economic decision criteria.

Following are brief comments on the most significant constraints that seem to be becoming institutionalized into the NFMA planning process. These, surprisingly, are not environmental constraints designed to generate nonmonetary benefits. Environmental constraints can be dealt with by the cost-benefit analysis methodology outlined above. The most significant constraints have actually evolved from timber management concepts that were developed by the forestry profession long before "environment" become a household word. Nondeclining yield is by far the most significant of these constraints. Related constraints include rotation ages that maximize mean annual increment, ending inventory, and management prescriptions that will affect the amount of land determined to be unsuitable for timber production.

The Forest Service cites both the NFMA and its own policy of managing the national forests for sawtimber-size trees as the basis for its regulation requiring stands to be grown to an age approaching culmination of mean annual increment (CMAI). There is a trivial amount of leeway allowed, since the stands need reach only 95 percent of CMAI. Most of the existing timber on the national forest is already so old that it is considerably past CMAI. By itself this regulation would have relatively little economic impact, as several decades will pass before the age of the oldest remaining stands left to harvest drops to CMAI. However, when this regulation is coupled with the nondeclining evenflow regulation and both are treated as rigid constraints in FORPLAN, the results are in considerable conflict with the objective of maximizing the net benefits to society from the forest. From an economic point of view, specification of any fixed rotation age, even an economically efficient one, is an unnecessary constraint. The model should be free to select stands for harvest at any age in order to maximize benefits. The rotation age is simultaneously determined along with optimal harvest levels for each decade.

This leads to another problem with rotation ages based on CMAI. When intensive management regimes are selected to maximize the present value of a forest, they often have an age of CMAI that is greater than for natural (unmanaged) stands. Strictly following the CMAI regulation could ultimately result in lower harvest levels and net benefits than if the land was not intensively managed.

CMAI need not be totally in conflict with economic efficiency if the flexibility permitted in the Forest Service regulations is applied where appropriate. The regulations specifically allow regional plans to permit harvesting at ages below CMAI for periods as long as thirty years in cases "where missing age classes cause disruptions in nondeclining flow which can be smoothed by harvesting trees prior to culmination of mean annual increment. . ." (36 C.F.R. 219.12(d)(ii)(C)). Several other exceptions to CMAI are also permitted in regional plans. In addition, the region can exercise considerable control over the selection of the age at which culmination occurs during the process of developing managed yield tables.

However, it remains to be seen if regional plans will take advantage of the flexibility allowed under the regulations, given the unjustified Forest Service objective of producing sawtimber rather than a mix of pulpwood and sawtimber. Further, it is clear at this point that the agency has developed the regulation to support its traditional rotation age policies, rather than incorporating meaningful economic efficiency criteria.

On more than one occasion I have been told by Forest Service planners that certain constraints are required by law and that it would be illegal to make computer runs without them. They may have been pulling my leg, but I really do not think so. Further discussions sometimes led to an admission that there is nothing illegal about making any kind of computer run. Unfortunately, there is still a widespread belief that all the formal alternatives included in an EIS must conform to existing laws and policies. The regulations implementing the National Environmental Policy Act of 1969 (NEPA) are quite clear on this point. They specifically state that "for the purpose of sharply defining the issues and providing a clear basis of choice among options by the decisionmaker and the public," agencies are expected to "include reasonable alternatives not within the jurisdiction of the agency" (42 U.S.C.A. 4332(2)(D)).

The process, described earlier, of starting with an unconstrained benchmark run would automatically result in departures from nondeclining yield as well as long-run equilibrium harvest ages based on financial maturity rather than on culmination of mean annual increment. Economically optimal timber management prescriptions would also ensure that once an area became economically accessible for logging, the residual stand or cutover land would continue to be suitable for timber production. Practices that would not contribute to a positive net present value would simply be excluded in the solution of an unconstrained run. The opportunity costs of imposing these constraints and their interaction with other constraints would then become visible and would provide a basis for accepting or rejecting them as part of the final plan. If laws need to be changed before the Forest Service would adopt a particular plan that appears to be in the public interest, Congress would at least have some good analyses to help make more intelligent laws.

The Choice of An Interest Rate

Interim Directive No. 7, <u>Economic and Social Analysis</u>, states as follows: "For the 1980-1985 long-term National Forest and regional planning processes, use a 4 percent real discount rate. In addition, use a 7-1/8 percent real discount rate to determine the sensitivity of alternatives to variations in the discount rate [the 7-1/8 percent rate was used in the 1980 RPA]" (USDA, Forest Service, 1980f). The Forest Service had earlier submitted a document to OMB to justify a 4 percent rate rather than a 10 percent rate, which OMB has directed all governmental agencies to use for evaluating government projects unless they were covered by a Water Resources Council directive that provided the 7-1/8 percent rate. (Interim Directive No. 7 is another example of the Forest Service autonomy discussed earlier.)

The choice of the interest rate for evaluating government projects is one of the most important decisions affecting the level of public investment activity. This is particularly true for forestry, because some investments require waiting periods of several decades before any returns are realized. A small change in the rate of interest can significantly affect the economically optimal way that forests should be managed. Also at stake in the choice of the proper interest rate is the allocation of resources between the public and private sectors of the economy. A relatively low rate will lead to a higher proportion of the economy's activity being operated by government agencies. A relatively high interest rate will leave more economic activity in the hands of private enterprise.

There are three basic approaches used to select an interest rate for government projects. These are: (1) the social rate of time preference; (2) the cost to the Treasury of borrowing; and (3) the opportunity cost of displaced private spending. Each approach has its advocates, but the opportunity cost approach has by far the strongest support from the economics profession. In fact, in 1968 the Subcommittee on Economy in Government of the Joint Economic Committee of the U.S. Congress, after extensive hearings, recommended "that no public investment be deemed 'economic' or 'efficient' if it fails to yield overall benefits which are at least as great as those which the same resources would have produced if left in the private sector." The subcommittee found that "currently, the rate of return on alternative minimum-risk private spending is at least 5 percent" and that "explicit allowances be made for risk and uncertainty in the benefit and cost estimates of each public investment" (U.S. Congress, 1968).

Some of the economists that appeared before the subcommittee argued for substantially higher interest rates--rates in the 7 to 12 percent range. In his concluding remark to the subcommittee, William J. Baumol, Professor of Economics at Princeton University and a leading authority in in this field, stated that

> while there is not complete unanimity among economists on the precise number that should be used in discounting it would be misleading to infer that there is any disagreement on the basic point at issue. The profession speaks with one voice in asserting that a discount figure of 3.5 to 4 percent is too low in present circumstances, and warns us clearly of misallocation of resources and inefficiencies that are likely to result from the use of such unjustifiable figures (U.S. Congress, 1967).

Since these hearings, there have been several additions to the literature in this area, but no substantive arguments have been made to refute the subcommittee's or Baumol's findings and recommendations. There have been a significant decrease in the rate of growth of the economy and much higher rates of inflation. The lower growth rate of the economy has somewhat reduced the opportunity cost of displaced private spending, but this has been more than offset by the increased premium for risk and uncertainty that must be added to the risk-free rate of return resulting from higher rates of inflation.

The interest rate is extremely important in forest management if it is used correctly to determine optimal management practices. Such practices would be designed to maximize the net present value of the projected streams of costs and benefits for the specific rate of interests used in the discounting calculations. A change in the discount rate will affect the optimal practices. Simply calculating present values of the same practices with different interest rates is an improper technique. Economically optimal forest management plans would use the mix of management practices that equates the marginal cost and price of the stumpage over time. The values of each marginal acre in each type of land use would also be identical. These are necessary conditions for management plans that will maximize the net present value of the projected streams of costs and benefits from all forest uses.

When economic analyses are done to analyze limited aspects of a forest management plan based primarily on traditional forestry criteria, it is very difficult to make unambiguous statements about the effect of interest rates. The use of either noneconomic criteria or constraints or both can make some results completely insensitive to the interest rate. Other results can move in either the same direction or in the opposite direction from what they would in economically optimal forest management plans. No general rules can be applied. It is necessary to specify completely the exact model being used before unambiguous statements can be made about the effect of interest rates.

There is currently a very active debate going on within industry circles about whether or not to support the Forest Service's 4 percent rate or a higher rate such as 6 or 8 percent. I believe the outcome of this debate will be to insist that the Forest Service use at least two rates, say the 4 percent and the 7-1/8 percent, as provided in Interim Directive No. 7, with the proviso that plans be optimized for each of these rates. Several Forest Service planners have said they will only make constrained optimization runs at 4 percent and then simply rediscount the net benefits of the recommended plan at 7-1/8 percent.

"Single-Acre" Analyses

Two general types of analysis are frequently made to determine the economic impact of modifying various management criteria or constraints or partially to use economics to help specify some practices. These can be called the "single-acre" approach and the allowable cut effect (ACE) approach. The former is to determine economically optimal management practices for a single acre, using the Faustmann formula to calculate maximum soil expectation values for a given rate of interest. This approach can properly determine the economically optimal mix of management practices, including rotation ages, for a single acre of land or for a large number of acres, provided that the price assumptions that are used for such analyses remain valid when the indicated optimal mix of management practices is carried out on every acre.

With the imbalanced age class distribution on many large forest properties, including virtually all the national forests, the results of single-acre analysis are seldom used without some modification or limitation. One typical modification is to use single-acre analysis to obtain some input assumptions for what is otherwise a traditional approach to allowable-cut calculations. This is frequently done where a single-acre analysis indicates that a large block or even a whole forest of old-growth should be harvested immediately. The allowable-cut calculations can be made using traditional textbook formulas, or with any of the more recent computer simulation or linear programming models such as FORPLAN. The input assumptions derived from single-acre analysis can include silvicultural treatments and rotation ages. These assumptions can apply to existing timber stands that will be regenerated in the near future and to stands that will not be obtained for many decades but that are used to define a target forest.

When single-acre management assumptions are input to an allowable-cut calculation, some very perverse results can be obtained. These results are dominated by the type of harvest flow constraints used in the calculations.

The economic test for lands suitable for timber production under the NFMA regulations is a variant of the single-acre approach, but it is still unclear whether or not an economically optimal management regime will be used for the test. If the management regimes or rotation ages used for this test are not economically optimal, the results are indeterminate. With economically optimal management, no acre should have a negative net present value and, consequently, no acre should be declared unsuitable for timber production. Any practice that contributes to a negative net present value would simply be excluded. If management practices such as regeneration prescriptions or rotation ages are imposed as constraints before the economic test is made, then the effect of the interest rate on a national forest management plan is indeterminate.

Nonoptimal management practices will cause those acres closest to the margin to become submarginal. How this affects a total forest management plan will depend upon whether or not the management constraints imposed on those acres that remain supramarginal increase wood production sufficiently to offset the creation of submarginal acres. Once a set of noneconomic constraints is imposed on a forest management plan, however, the higher the interest rate, the larger the number of acres that will be classed as submarginal.

Allowable Cut Effect Planning

The second major approach to mixing economic analysis and traditional forest management planning incorporates what has become known as the allowable cut effect. With an evenflow harvest constraint, any silvicultural practice or land use decision that raises or lowers the long-run sustained yield level of a timber management plan results in a corresponding increase or decrease in the current allowable cut, provided there is sufficient mature inventory available to maintain the new harvest level indefinitely.

Harvest flow constraints frequently eliminate or severely limit the effects of most economic variables that should be used in forest planning. In fact, past planning practices with Timber RAM have even made the so-called objective functions irrelevant. FORPLAN is capable of producing the same kind of results where harvest levels are the same regardless of whether the objective is defined to be maximum timber production or maximum present net worth. The harvest flow constraints and their interaction with other noneconomic variables completely dominate the solution.

Evenflow constraints give rise to numerous anomalies. Practices that are clearly uneconomic on a single-acre basis, but which will increase long-run sustained yield levels, may appear to be economic because of their allowable cut effect. The increase in present value from accelerating current harvests more than offsets the reduction in present value from uneconomic practices that create an allowable cut effect. The higher the interest rate, the greater the number of uneconomic practices associated with a so-called target forest that can be offset by accelerated harvesting of existing inventory.

Practitioners of the allowable-cut approach to forest planning normally rely on ad hoc simulation or constrained optimization techniques that are extremely varied. Attempts to include all forestry practices that could produce an offsetting allowable cut effect are usually limited by a budget constraint, by the planners' intuition, or simply by tradition. Logically consistent economic analyses and conclusions are impossible. The effect of various interest rates depends solely on the inclinations of the various practitioners of allowable-cut calculations and on how they chose to let interest rates and any other economic variables affect their calculations. All economic calculations associated with this approach are simply used as a veneer to obfuscate the real planning criteria or to comply with some regulation or directive. The rate of interest is really irrelevant unless the practitioners want to make it appear otherwise.

In addition to the interest rate, other price or benefit assumptions can be extremely important in forest planning. These other price assumptions will affect the reported present values for all the planning approaches. They can also affect the management prescriptions used in all the planning approaches except for the traditional forest management approach, which totally ignores prices. Only the economically optimal approach will optimize with respect to prices, however. The mixing of traditional forest planning with some economic calculations invariably leads to more anomalies.

When economically optimal planning is done, it is very important to make reasonable price projections. These projections must take into account whether or not the planning unit can affect either production input or output prices or both by the amount of timber it can sell or by varying any other management activity. If the planning unit can affect prices, economically optimal plans can be obtained only by using calculation techniques that properly account for price changes when the level of activity changes. If the planning unit cannot affect prices, then any calculation technique that uses harvest flow or any other activity constraints will not produce economically optimal plans.

The Economic Objective Function

The final topic to be discussed here is the economic objective function that should be maximized for NFMA planning. When forests have downward sloping curves, the forest has the option of using its monopolistic powers to maximize net cash flow to the U.S. Treasury and in lieu payments to the affected counties. This would mean restricting timber harvests, for example, to the point where marginal revenue equals marginal costs. Marginal user cost as defined by Scott (1953) is implicitly included in marginal costs. FORPLAN has this monopolistic, or more correctly, oligopolistic objective function as an option that is referred to in the draft FORPLAN manual and other Forest Service material as maximizing present net worth.

The other economic objective function that FORPLAN can maximize when it functions properly is the maximization of present net benefit. This is defined as the areas under the demand curves and above the marginal cost curves for timber. Present net benefits are maximized by jointly maximizing the present value of the net cash flows obtained by the Forest Service _and_ the timber purchasers. To the extent that individual timber purchasers buy their timber and sell their products in competitive markets, any benefits they receive from lower stumpage prices will be passed on to their customers. Society's net benefits are maximized by timber output levels that equate supply and demand or, to be more precise, marginal costs and prices.

Because maximization of present net benefits produces the levels of output and prices associated with competitive markets, I believe this is the correct economic objective function for the Forest Service. Plans that maximize present net benefits always result in higher resource outputs and lower prices than oligopolistic plans. In the work that I have done to date comparing the results of these two economic objective functions, I have found that Forest Service timber outputs have been consistently less than the oligopolistic optimum. In other words, increasing timber outputs slightly would actually increase returns to the Treasury. After the point of maximum return to the Treasury was reached, the additional output required to achieve present net benefit maximization would reduce the returns to the Treasury.

You may recall the following, quoted earlier from the draft economics handbook: "Explicit recognition of market dominance and inelastic demand raises the real question of pricing strategy--an approach that is rejected out-of-hand since the Forest Service does not operate as a profit-maximizing producer" (USDA, Forest Service, 1980b). No one to my knowledge expects or wants the Forest Service to act as a profit maximizer, but it is imperative that the agency be aware of and explicitly account for its impact on prices. The objective is not to have the Forest Service act like a profit-maximizing oligopolist, but to produce forest outputs consistent with the outcome of competitive markets. Instead, we have a situation in which we as taxpayers and consumers would be better off if the national forests were simply given to an Arab shiek. An OPEC-type cartel would find it profitable to sell more timber at lower prices that would benefit consumers. In addition, the taxpayers would not have to pay the $1 billion per year subsidy now needed for the Forest Service. This, however, would still fall far short of what we should expect from the Forest Service. Even more timber and lower prices, as well as positive net returns to the Treasury, should be achieved if the NFMA planning process will lead to the highly desirable, yet elusive goal of maximizing net social benefits.

CONCLUSIONS

In conclusion, I am in substantial agreement with every point except one made by Krutilla, Bowes, and Wilman in chapter 6 of this volume. I agree fully with the planning model they have presented, and believe it corresponds very closely to what is presented in this chapter. The one area of disagreement is the extent of the movement of the Forest Service in the direction that both chapters agree it should go. If it is moving, it is at such a glacial pace that I do not perceive any movement. The draft manuals and other material appearing now do not reflect the level of understanding needed to make the necessary changes, and traditional concepts are still being clung to tenaciously. The change that Krutilla, Bowes, and Wilman perceive may become meaningful by the time there is a 1995 or 2000 RPA Program, but I do not think we can wait that long. If the changes are not very apparent in the first NFMA forest plans prepared prior to September 1985, with perhaps some slippage in that schedule, I do not think RPA and NFMA will still be around in 1995 or 2000, but that instead the sagebrush rebellion will have spread to the timber, and the Forest Service as we know it today will cease to exist.

REFERENCES

Benninghoff, B., and coauthors. 1980. *Region 2 Timber Demand Analysis*. Draft Report to Region 2 Planning Team. July 1980 (USDA, Forest Service).

Congressional Record. 1980. Vol. 126, no. 159, 96 Cong. 2 sess., pp. 14462-63.

Federal Register Office. 1979. "Statement: Support for the President's Energy Program," in *Weekly Compilation of Presidential Documents* vol. 15 (July-September) (Washington, D.C.).

Krutilla, J. V., and J. A. Haigh. 1978. "An Integrated Approach to National Forest Management," *Environmental Law* vol. 8, no. 2.

Scott, A. 1953. "Notes on User Costs," *Economic Journal* (June), pp. 368-384.

U.S. Congress. 1967. *Hearings Before the Subcommittee on Economy in Government of the Joint Economic Committee*, 90 Cong. 1 sess. (Washington, D.C., GPO), pp. 152-159.

_____. 1968. *Economic Analysis of Public Investment Decisions: Interest Rate Policy and Discounting Analysis*. Report of the Subcommittee on Economy in Government of the Joint Economic Committee, 90 Cong. 2 sess., September 23, 1968 (Washington, D.C., GPO).

USDA, Forest Service. 1980a. *An Assessment of the Forest and Rangeland Situation in the United States*. FS 345 (Washington, D.C.).

_____. 1980b. *Economic Analysis for Forest Planning--Review Draft*. Forest Service Handbook, 1909.12, cpt. 560 (Washington, D.C.).

_____. 1980c. *A Recommended Renewable Resources Program*. FS 346 (Washington, D.C.).

_____. 1980d. *The Proposed Lolo National Forest Plan*. April 1980 (Missoula, Mont., Lolo National Forest).

_____. 1980e. *The Lolo National Forest Plan Draft Environmental Impact Statement*. April 11, 1980 (Missoula, Mont., Lolo National Forest).

_____. 1980f. *Economic and Social Analysis*. Interim Directive No. 7. USFS Manual, cpt. 1970. September 2, 1980 (Washington, D.C.).

_____. 1980g. *Draft Environmental Impact Statement--The Northern Region Plan*. November 17, 1980 (Missoula, Mont., Northern Region).

_____. 1980h. *The Proposed Northern Region Plan*. November 17, 1980 (Missoula, Mont., Northern Region).

Walker, John. 1977. "Economic Efficiency and the National Forest Management Act of 1976," *Journal of Forestry* (November).

West Virginia Division of the Izaak Walton League v. Butz. 1975. 522 F.2d 945 (4th Cir.).

David A. Anderson

COMMENTS

I can appreciate some of John Walker's frustration concerning the status of specific analysis procedures for preparing forest plans. The Forest Service is simultaneously developing procedures and producing plans. This, combined with the fact that the state of the art has not advanced as far as we would like, has led to a confusion on specific policies and analysis procedures.

Relative to Walter's comments on economic efficiency and economically optimal solutions, I agree with the concept that allocation of resources ideally should maximize net contribution to social welfare. To do this, there would need to be agreement--by all members of society--on downward-sloping demand curves for all outputs and effects associated with the management of a national forest. With such demand curves, there would be no need for constraints--or even alternatives for that matter. When these demand curves were equated with the supply curves in a linear programming model, the perfect solution would emerge. In that case, everyone would have agreed to the demand curves, everyone would be satisfied with the answer, and the plan could be implemented. However, to my knowledge, nobody has been able to develop demand curves for all goods and services. And, as discussed by various authors in this volume, the data for such an approach do not exist.

In the Rocky Mountain region we have attempted to develop a downward-sloping demand curve for timber and additional demands curves for the grazing and water resources. We seem to have agreement on the timber, but are far from agreement on grazing and water. Apparently, the state of the art is just not equal to the task.

The Committee of Scientists, in their final report relating to the Forest Service regulations, said: "Analysis for determination of both efficiency and impacts has generated considerable debate. Much of it centers on the 'state of the art' and the possibilities of a given technique being universally practical for nationwide implementation" (44 Fed. Reg. 53,951 (1979)).

This very issue was instrumental in the use of the term "cost efficient" in the National Forest Management Act (NFMA) regulations. The term "economic efficiency" seemed to indicate that there was an all-inclusive set of demand and supply curves and that the economically efficient solution could be found. The term "cost efficient" seemed to reflect the state of the art.

For those outputs having generally agreed-upon demand curves, we intend to use the net present benefit generated from the curves to allocate resources. However, for those outputs and effects having no demand curves, we are forced to use constraints. Cost efficiency does require that the constraints be met in a cost-efficient manner.

I mentioned the demand curves that we developed for the Rocky Mountain region. In addition, we are developing and testing a downward-sloping demand curve for timber on the Black Hills national forests. Preliminary tests indicate no significant difference in results between the horizontal and downward-sloping demand curve. We will continue to test the differences.

I agree with Walker that constraints and demand curves should be analyzed--and that the decision makers, the public, and Congress should know the consequences of their decisions. Current procedures direct each national forest to develop a broad range of prescriptions from which a broad range of alternatives can be analyzed. We are following the Council of Environmental Quality regulations which instruct us to consider alternatives outside the jurisdiction of the agency. An "unconstrained maximum" alternative is required by all national forests in the Rocky Mountain region--for at least two resources, including timber.

The basic position of the Forest Service has been and continues to be a willingness to analyze. If there are agreed-upon demand curves, we will use them. However, I do not think the Forest Service should start from the premise of trying to define all conceivable alternatives. This would not be cost-efficient planning.

Concerning the discount rate, I agree that the opportunity costs of diverting resources from private use are the traditional and most accepted rationale for setting the federal discount rate. In practice, however, problems occur in measuring this rate, especially for the long-term planning being done in the Forest Service.

Paul W. Boltz, a senior economist with the Board of Governors of the Federal Reserve System, after a study of AAA corporate bonds from 1960 to 1978, stated that, "Given the productivity of physical data, the real rate of return would not be expected to depart much from 2 1/2 percent or so for long periods of time. . . . It appears that the real rate of return on corporate capital is now around 2 1/4 to 2 1/3 percent" (Paul W. Boltz, letter to Fred Kaiser, Forest Service economist, April 17, 1979).

Since the private sector must pay tax and the public sector does not, the discount rate for the public sector should be adjusted to reflect this tax. The average corporate tax rate is about 40 percent. Therefore, with a 2 1/4 percent real rate of return and a 40 percent tax adjustment, the real discount rate for the public sector should be about 4 percent. The Forest Service directs its economists and planners to analyze the discount rate. They are instructed to run an entire allocation at 7 1/8 percent and not to just recalculate a 4 percent allocation at 7 1/8 percent.

Regarding objective functions, there has been some confusion over which objective to use when analyzing with a downward-sloping demand function. The FORPLAN model has two methods of maximizing the objective function. One is called net present worth (NPW). With this objective function, the model will maximize revenue to the government. It will make the producer act like a monopolist, and reduce production on timber stumpage to raise the price. This objective function stops production where marginal cost equals marginal revenue. All national forests have been instructed not to use the NPW objective function. The government must take both producer and consumer points of view.

The second objective function is net present benefit (NPB). This objective function stops production where marginal cost equals price. It is my understanding that this objective function maximizes both producer and consumer surplus, and so maximizes net social welfare. Each forest has been instructed to use this objective function if it has a downward-sloping demand function.

I believe it is fair to say that the Forest Service knows how to do the proper calculations. If our models are not doing them correctly, we will correct them. Also, we recognize that since there is no all-inclusive set of demand curves, we will be doing a partial analysis at best.

Let me add an observation on simulation and optimization. It is my contention that, because we are dealing in the social sciences with such concepts as economic efficiency, the perfect model will never be built. The best we can do, then, is to use the optimization tools in a simulation mode and to make the results of the analyses available to the managers.

The Forest Service is going through a transition relative to strategic planning. A significant amount of retooling is going on in the areas of new skills, data and information management, analytical procedures, and computer systems application. Because the Forest Service is faced with process development and production at the same time, some inefficiencies are necessarily going to be encountered. At the same time, professional judgments are going to have to be made. Decisions are made every day on the ground. Unless there are some major conceptual differences in procedure, I believe it is in the best interest of everyone to finish the first round of planning.